BIGFOOT

What we know 2025

A Field Researcher's Investigation into North America's Most Enduring Mystery

By
Timothy D.

Cover design by Timothy D

Interior layout by Timothy D

Printed in Usa, Canada

First Edition

ISBN: 978-1-0696715-1-6

Table Of Contents

Introduction

I've spent years walking logging roads, crouching behind cedar thickets, placing trail cams on old game runs, and sleeping alone in remote Northern Ontario bush. Not chasing legends—following patterns. Listening to voices that don't usually make it into books. Sorting stories from the noise.

Because out here, stories are all we have. Until they're not.

This book isn't about proving Bigfoot exists. If you need proof, you're already asking the wrong question. This is a field guide to what we've gathered, compared, discarded, and come back to again over nearly seventy years of boots-on-ground searching. It's a portrait, not a verdict.

I don't believe Bigfoot is a ghost, a spirit, or a traveler between dimensions. I don't believe in "mindspeak" or that it builds stick tepees to trade marbles. I believe it's an animal. A real, physical creature. Intelligent, adapted, and incredibly good at staying hidden—so good, in fact, it's turned the idea of proof into something almost mythical.

But the evidence is there. Scattered. Fragmented. Often ignored.

Prints. Sounds. Sightings. Behavior.

And in 2025, we have tools we've never had before. We have AI-enhanced analysis, multi-sensor drones, long-term autonomous audio units, and an international web of researchers sharing data in real time. This is not the same Bigfoot search from the 1970s. It's sharper. Quieter. And in some ways, a lot more unsettling.

This book is structured not to convince, but to document. To compile what we know, what we've seen, what science can test, and what the field continues to quietly whisper from the edge of the dark. Yes, it's largely focused on North America—because that's where I live, and it's where the data still stacks highest. But similar patterns, similar beings, similar voices emerge across the globe. Those are not coincidences. They are convergences.

You'll find field reports in these pages—some mine, some from others. You'll find scientific analysis next to raw emotional testimony. You'll also find things that feel like they shouldn't be possible. Those are the moments worth leaning into.

I'm not here to sell folklore. But I respect its roots. And I believe that the truth—whatever it turns out to be—will come not from myth, but from the wilderness itself. From those late-night moments when the silence isn't quite silent. When something just outside your light stops moving the moment you do.

This is not fiction. It's a modern-day, real-life adventure.

And like all true wilderness journeys, it's the experience—the tracking, the waiting, the questions—that matters most.

Chapter 1: The Sightings Haven't Stopped

Part 1: What the Reports Still Say

There's a myth that this whole thing died in the seventies. That Bigfoot was just a fever dream of grainy footage, plaster casts, and late-night radio. That once the documentaries faded and the books stopped selling, the creature slipped into the same drawer as Elvis and the Loch Ness Monster. But that's not what's happening.

If anything, the reports are more consistent now than they've ever been. Not louder—just more precise. Less storytelling, more field notes. And the places these sightings come from… they aren't YouTube comment sections. They're ridge trails, hunting blinds, fire roads, cold lakes before dawn. They're mine.

Northern Ontario doesn't make national headlines. It never did. Too quiet. Too remote. Too indifferent to the noise of the south. But if you know where to listen—really listen—you start to realize it's one of the most active encounter zones on the continent. Places like the Kenora District. Capreol. Rabbit Lake. Marten River. These aren't one-off stories. These are zones. Places where people still talk—quietly—about seeing things they wish they hadn't.

I've talked to those people. Loggers. Trappers. Firefighters. Quiet men who don't like cameras, and like attention even less. They don't tell ghost stories. They describe weight. Movement. Muscle. Gait. They describe something real. Something they can't file under moose or bear. And they describe it with a kind of hesitation I've come to recognize—almost like confession. Not fear. Just a heavy kind of knowing.

Let me give you one of the first ones that stuck. Fort Hope. Summer of 2022. Middle of the day. A man driving alone on a secondary road saw two figures cross ahead of him—large, upright, fast. He didn't say "Bigfoot" at first. He just said, "They moved like people, but they weren't people." No bobbing head. Arms too long. Disappeared into the treeline so fast it made the silence behind them feel wrong. A few days later, a group came through the same area on ATVs. Found tracks. Deep ones. Good toe splay. Castable. A thermal drone team swept the area and picked up heat signatures that didn't match deer or moose—too upright, too fast—but didn't hold long enough for visual confirmation. That case is still open in my log. And it's not an outlier.

I've documented over a dozen credible Ontario encounters in just the last decade. Not the ones you find online. The ones people only talk about when no one else is around. When they think you'll believe them. When they think maybe someone else saw it too. And here's what stays the same every single time: remote location, large upright figure, not a known species, vanishes before evidence can catch up. From Madawaska to Moose Factory, the details stay locked. Not just in content—but in tone. These witnesses aren't excited. They're disturbed. They don't want attention. They want to feel sane again. And some of them never step back into the woods.

I had my own moment like that. Rabbit Lake, 2014. Paddling in solo, morning glass on the water, fog holding low to the trees. No birds. No wind. Just breath and paddle. I wasn't thinking about anything paranormal—I was scouting a new camp site. Then came the sound. One long, rising howl. Too deep for wolf. No bubble like moose. It was one solid frequency that cut and dropped off like someone let go of a rope. I drifted for a long time after that. Didn't land right away. When I did, it was careful. Slow. The bush felt heavy. Not dangerous—just waiting. I didn't find tracks. But I marked the direction, logged the sound, and wrote the moment down like I always do. I didn't write "Bigfoot." I didn't need to.

It's not about belief anymore. It's about patterns. And the pattern hasn't stopped.

Part 2: The Ontario Silence

People like to think that if something this strange were really out there, we'd have caught it by now. That a species this big and biologically distinct couldn't possibly evade every camera, every drone, every hunter with a GoPro. That in a world where phones track our heart rates and satellites can read license plates, something like Bigfoot would have been cornered, documented, pinned to a data set, and named. But that's only true if you understand how little of Northern Ontario gets seen—let alone searched.

Even an hour outside Sudbury, there are entire ranges of black spruce, glacial ridgelines, inland muskeg and boulder-choked valleys where no one goes unless they have a reason—and even then, only for a few months out of the year. There's no cell coverage. No roads. No mapped trails. It's not just hard to get to. It's hard to survive in. Add in spring runoff, fall storms, blackflies so thick you breathe them, and terrain that eats boots and breaks gear. Now imagine trying to search that systematically. You can't. And that's why the silence around this phenomenon isn't suspicious. It's expected.

Ontario is over a million square kilometers. Larger than most countries. And most of its northern half is forest, wetland, rock and water. The deeper you go, the fewer people you see—and the more likely you are to hear something no one else will. That's where the good sightings come from. Not the tourist routes. The logging roads that don't exist on maps. The portage trails used once a decade. The high ground where caribou pass and humans don't.

And when you get a report from one of those places, it doesn't come through an app or a hotline. It comes from a neighbor. A friend of a friend. A logger who didn't want to talk, but finally said something. A cousin who grew up in the bush and came back different. These people don't follow the research. Some don't have email. Some don't have phones. When they do talk, it's careful. Quiet. And full of details no hoaxer would think to include. Things like direction of movement. Shoulder width. The way the head didn't bob like a human's. How everything went dead quiet before and after.

I'll give you one. 2015. Capreol, northeast of Sudbury. A group of hikers—locals, not tourists— saw something tall, dark, and still just inside the tree line near a watering hole. They froze. The thing didn't move. They said it wasn't a bear. They said it wasn't a moose. It stood upright. No snorting. No sound. After maybe ten seconds, it turned and walked—walked—deeper into the woods. No crashing. Just movement. No one pulled out a phone. They were too stunned.

That story might've ended there. But months later, another group came through the same trail. This time two turkey hunters. Same area. Same watering hole. Same description. No knowledge of the first report. I found them both in follow-ups and matched their stories word for word. That kind of convergence doesn't happen by accident.

Same thing in Sioux Narrows. Multiple reports in 2006 and 2009 around the same cabins. One involved a bear carcass that went missing—lifted, dragged, no blood trail. The summer workers who were there told me it wasn't just the theft. It was the feeling of being watched in the hours before. Same with Longlac in 2012—feces found near a trail, tracks in soft sand, and one hunter who described the smell like "damp wool and blood."

And these aren't stories from decades ago. I'm not quoting your grandfather's neighbor's uncle. These are recent. Names I've met. GPS coordinates I've stood on. Photos I've held in hand. And every time I collect one, I think about how many didn't get told.

Because that's the other silence. The human one.

Even now, most people don't report what they see. Not to police. Not to databases. Not even to their families. Because once you say the word—once you tell someone you saw a "Bigfoot"—you're not someone who saw something anymore. You're someone who believes something. And that's when the distance sets in. From your coworkers. From your friends. Sometimes even from your own voice, because once you've said it out loud, you can't unsay it. So people shut up. They go quiet. They hold the memory like a wound.

But those memories don't go away. And every once in a while, they come out.

When they do, they sound a lot like mine.

Part 3: Ground Truth

There's a point in this work where something shifts. Not in the forest—in you. It's after enough cold nights, enough bad batteries, enough hours sitting in the dark waiting for something that doesn't show itself. It's after hearing one too many quiet, nearly identical stories from people who've never met. And when that shift comes, you stop asking *if* the phenomenon is real. You start asking *why no one wants to deal with it*.

I hit that point years ago.

It wasn't a single encounter that pushed me over. It was accumulation. Pattern. The quiet stacking of stories from people with nothing to gain. They didn't want to be believed. They wanted to stop remembering. But they remembered anyway, because what they saw didn't go away. It sat there, just beneath the surface of their lives, waiting for someone like me to ask. And when I did, the details lined up. Not vaguely. Not approximately. *Exactly.* That's what field researchers call ground truth. The point where the noise drops out and only the consistent data remains.

And here's what's consistent: people are seeing something upright, large, and bipedal. And they've been seeing it for over a hundred years.

In Ontario alone, we have reports dating back to 1906—the so-called "Old Yellow Top" in Cobalt—and they haven't stopped. I've logged cases from Thunder Bay to the Sault, from North Bay to Temagami, from Rabbit Lake to Cochrane. Every decade. Every season. Same description. Same result: the creature vanishes into the bush before anyone can react.

Take Killarney, 2015. A group of residents heard screaming in the early hours. Not moose. Not human. Something raw and vocalizing from just beyond the clearing. That same day, a tall figure was spotted running along an old service road. Not crashing through brush—moving fast, quiet, like it knew where to go. One of the witnesses was a nurse. Didn't believe in any of this. Said it "sounded like a man screaming, but the chest was too big. Too deep. Like a voice too large to be from a throat."

Now put that next to a 1992 patrol report from the Cochrane District. A police officer, driving alone, sees a massive upright figure cross the Trans-Canada in two strides. Caught in the headlights for less than a second. His notes say, "Too big for a bear. Too fast for a person." Then it was gone.

Or go back to 1987, Sioux Lookout. A trapper on his line. Middle of the day. Face to face with something that looked like a man—but wasn't. Said it stared. Then turned and walked off without a sound. He never told anyone but his son. That notebook only surfaced after the trapper passed.

Or more recently—2013, Mowat area. A family staying in a remote cabin reported continued interactions with what they believe was a *group* of creatures. Rocks thrown, whoops at night, shadows on the edge of the treeline. No interest in going public. But they still won't go back.

Different years. Different people. Same animal.

Even outside Ontario, the pattern doesn't change. In BC, Bigfoot sightings are practically seasonal. In the Cascades, backcountry rangers quietly keep lists. In Kentucky and the Smokies, reports rise after forest fires or logging—like something's being pushed out. And across the eastern provinces, from Nova Scotia to Quebec, I've heard the same phrase whispered by fishermen and hunters: "It wasn't a bear."

And the most telling details? They're not dramatic. They're subtle. The weight in a footstep. The way birds go silent. The width of eye shine that's too far apart to be a deer. The broken branch that shouldn't be there. The pressure in your chest when you know something's watching you— but you can't find it.

I've felt that. Many times.

There's a hill just west of Wawa where the wind dies fast, like something turned off the sky. You don't hear birds when you stand on that ridge. Just air. And silence. There's a game trail near Temagami where the forest floor is pressed flat for forty feet—but there's no sign of what made it. No claw marks. No antler rub. Just pressure and pattern. And there's a lakeshore in Noganosh where a local family has tracked sightings, knocks, and tracks for a decade—but they don't talk about it anymore. Not even to me.

That's the other kind of silence. Not absence of sound—but absence of *sharing*. A cultural hush around something people are seeing. Because Bigfoot became a meme. A mascot. A punchline. And when that happens, the real witnesses—the ones who felt it, smelled it, watched it walk away—they stop talking. Because they know what happens when you say it out loud.

But they haven't stopped seeing it. Not in 2025. Not even close.

If anything, more people are seeing them. Not believers. Not cryptid chasers. Just people who went into the woods and came back with a story they didn't ask for. People who started measuring stride lengths. Recording audio. Taking samples. Not to prove anything—but because they couldn't forget what happened. That's who I trust. Not the loudest voices. The quietest ones.

And they're the reason I'm still walking into the woods.

Chapter 2: The Pattern in the Descriptions

Part 1: Anatomy of a Match

You hear it enough times, from enough different people, and it starts to feel rehearsed. But it isn't. That's the disturbing part. These are not coordinated stories. They're isolated encounters from unrelated witnesses—people who've never met, never spoken, often don't even believe in Bigfoot themselves. And yet, the description they give is almost always the same.

Not similar. The same.

It starts with size. Seven to nine feet tall. Broad-shouldered. Upright. Walking on two legs. Covered in dark brown or black hair—sometimes reddish in the sun. Arms too long for the body. Gait too fluid for a bear. Head sometimes described as conical or sloped. No visible neck. And the movement—always the movement—silent, fast, purposeful. Not crashing like a moose. Not lumbering like a bear. Smooth.

You can find these descriptions in reports from British Columbia. From Pennsylvania. From the Ozarks. From the boreal forest north of Kenora. From an off-duty police officer who saw something standing in a clearing in Rainy River. From a retired schoolteacher who spotted a figure pacing a treeline near Red Gut Bay. From a couple who saw something walk upright through a cut line near Longlac. None of them were looking for it. None of them wanted to see it again.

And yet, when you place their stories side by side, they might as well be quoting each other.

I've studied the descriptions for over a decade—compared hundreds of field reports, personal interviews, and archival sightings going back to the 1950s. The language evolves a little, but the core traits do not. You could take a 1967 handwritten letter describing a "man-thing in the trees" and lay it next to a 2023 voice memo from a bowhunter in Sioux Lookout, and the details would line up like coordinates.

That's not mass delusion. That's biological consistency.

Even the wording tends to repeat. People say things like "the head was too low on the shoulders," or "the arms reached past the knees," or "it didn't move like a man—it moved like it didn't want to be seen." These phrases come up independently, again and again, across age groups, across education levels, across language. I've heard them from trappers, from campers, from Algonquin outfitters, from an Anishinaabe elder who'd never seen a television depiction of Sasquatch and simply called it, in Ojibwe, *Mishibizhiw-bineshiinh*—a quiet forest being. He didn't describe a spirit. He described a body.

And that's where it starts to shift. Because when you realize the consistency goes beyond shape and into *behavior*, it gets harder to dismiss.

They watch before they move. They freeze before they flee. They choose the tree line, never the open. They don't make noise unless they want you to hear it. They don't show up unless you're not ready. They don't run like deer. They vanish like something that understands how to.

And that's the word that comes up most often from people who've seen one up close. Not "monster." Not "creature." Not "spirit."

"Person."

"Big. Covered in hair. But a person."

I'm not saying they're human. I'm saying they behave like something that *thinks*.

Part 2: Faces, Eyes, and the Details That Shouldn't Align

It's easy to focus on size. Height, mass, stride length—it's measurable. It feels solid. But what makes people freeze isn't the size. It's the face.

Ask someone who's seen one. Really seen one, not a shadow or silhouette. Someone who looked close, even for a second. Ask them what they remember. They'll usually go quiet for a few moments. Then they say the same thing, almost every time.

"It looked back."

And when you press them—when you ask *what exactly* they saw—it always starts with the eyes.

They're described as deep-set, wide apart, sometimes amber, sometimes red in low light. Not glowing, not like something out of a movie, but reflecting—like a deer's, but not quite. Sometimes people say the pupils looked too round. Sometimes they say they looked human. But

what gets them every time is the expression. Not anger. Not confusion. Something closer to…
evaluation. Like it was measuring you back.

A man I interviewed outside of Marten River in 2018 put it this way: "It didn't seem scared of
me. But it also didn't want to be near me. Like it was deciding something."

He saw the face. Broad, with a wide, flat nose. Not gorilla-like, not exactly. More human in the
spacing, but with heavier brows and deeper skin folds. Skin visible around the eyes, and around
the cheeks where the hair thinned. Lips not as pronounced as a chimp—more like thin, dry skin
over a large jaw. And the eyes—he couldn't stop talking about the eyes.

That's what disturbed him most. Not the shape. Not the shock. The sense of *presence*.

You don't get that from a bear.

I've heard the same account from a hunter near Lake Temiskaming. From a woman whose dogs
refused to leave their tent for hours after she saw something squat near her camp, then stand up
and walk away. From a trucker who caught a face staring out of a tree line near Foleyet—just
staring. For ten, maybe fifteen seconds. Then it turned, stepped back into the trees, and
disappeared without a sound. He told me, "I've seen mountain lions. I've seen moose charge. But
that thing saw me. And didn't care."

Even more telling are the details that shouldn't line up—but do. The way the head sits too low on
the shoulders, making the neck seem absent. The elongated arms, with hands said to hang near
the knees. Fingers thick, but distinct. I've heard people talk about fingernails—dark, not claws.
And skin. Yes, they sometimes see skin. Around the face, the palms, the chest if the hair is
sparse. Dark grey, sometimes weathered, like callused leather.

These are not the details of someone inventing a monster. These are anatomical consistencies.
Repeated independently by strangers in different provinces and even different countries. They're
not describing a dream. They're describing something standing upright in front of them. And
that's what makes this so uncomfortable.

If it were a hallucination, the details would drift. If it were mass hysteria, the stories would
evolve with the culture. But they don't. They've stayed the same since long before the internet.
Long before Patterson-Gimlin. Before Sasquatch had a name that reached the public. These
people aren't describing Bigfoot. They're describing what they saw.

And what they saw looks back.

Part 3: Behavior That Doesn't Belong

The thing that catches people off guard—once the shock fades—is how the creature acts. Most witnesses expect chaos. A wild animal. Crashing, charging, grunting, maybe even aggression. But that's not what they see. They see control. Stillness. Decisions.

And that doesn't sit right.

Animals don't make decisions the way we do. They react. They flee, they bluff, they bolt. That's what people expect when they lock eyes with something that shouldn't be there. But over and over again, the people I speak to describe something else entirely. A moment of *pause*. A hesitation. As if the creature is assessing its options—*and them*.

There's a report from north of Chapleau, 2016. A pair of bowhunters in early season hiked deep off trail. Late afternoon. One moved ahead of the other to scout a ridge. He crested the slope and froze. Fifty yards away, standing at the edge of a clearing, was a tall, upright figure. It didn't run. It didn't snarl or move toward him. It simply turned its head. Looked at him. Then stepped backward into the trees, one foot at a time, until it vanished.

The hunter never raised his bow. He said it wasn't fear that stopped him. It was that he felt like he wasn't in charge of the encounter. Like the thing across from him had already made the decision, and he was just being allowed to watch.

Another witness, a trapper near Gogama, saw something massive crouched at a creek bank. Said it was scooping water, hands and all, no tools, no container—just dipping and drinking. It stood when he stepped into view. Not fast. Not startled. Just rose. Turned. And walked up the embankment. He never saw its face. Just the back. And how quiet it was.

Silence comes up constantly. These aren't encounters filled with broken limbs and growls. They're quiet. As if the creature has already learned that sound leads to danger. That stillness is a better strategy than escape. That humans only get dangerous when they know what they're looking at.

This kind of behavioral intelligence shouldn't come from an unclassified animal. It's not how moose behave. Or bears. It's how *we* behave when we're somewhere we shouldn't be—cautious, quiet, controlled.

Some researchers talk about this as a form of "avoidance intelligence." Not communication. Not culture. Just an evolved instinct to stay out of the line of sight. But it's more than that. It's the way these creatures avoid not just sight—but attention. They move at dusk. They avoid roads, even decommissioned ones. They don't come near camps unless they're watching. And if they leave signs—tracks, sounds, prints—they rarely leave them more than once.

And here's the unsettling part: they learn.

I've documented cases where a creature returned to the same site two or three nights in a row—until a camera was set up. Then nothing. No sound. No prints. No return. Not just avoidance. *Awareness*. It knew what changed. That pattern doesn't just suggest intelligence. It suggests memory. Feedback.

There's a zone near Noganosh where I set up passive recorders and trail cams over the course of five weeks. One recorder picked up a distant, rhythmic knock pattern—three knocks spaced wide, then silence. It repeated two nights later. The third night, after I repositioned a unit closer to the sound's origin, it stopped entirely. The camera was never triggered. But the sound never came back.

That's not random wildlife.

Even animals like wolves, bears, and lynx will often repeat paths or return to familiar ground unless there's danger. But this… whatever this is… seems to have a higher threshold for change. It monitors first. Reacts second. Just like we do.

It doesn't behave like a beast. It behaves like something *that knows how to behave around us*.

And that might be the most disturbing pattern of all.

Part 4: The Language of Consistency

If you strip away the fear, the doubt, and the cultural noise, what you're left with is language. Patterns of words. Repetitions in phrasing. And that's where things start to get clinical—because once you begin to map how people describe their experiences, something unusual appears.

People who've never met, in different provinces, sometimes decades apart, use the exact same words to explain what they saw.

Not similar ideas. *Identical phrasing*.

"He was huge—bigger than any man I've ever seen."
"The arms hung too low."
"It moved like it had somewhere to be."
"The eyes were deep, intelligent, like it was reading me."
"There was no sound. Like the forest had gone dead."

These aren't dramatic statements. They aren't literary. They're blunt, confused, sometimes whispered out of embarrassment. But I've heard each of those lines almost verbatim, in multiple interviews, from people separated by hundreds of kilometers and zero online connection. You don't get that kind of linguistic alignment unless something is being observed consistently.

In the field, we call this *witness convergence*. And it's one of the strongest indicators that a phenomenon is real. The more isolated the witnesses are from each other, the more weight their shared language carries. If twenty people describe something in the same way without influence, the odds of a coordinated hoax collapse. What you're left with is the uncomfortable possibility: they're all describing the same real thing.

And it doesn't end with Ontario.

I've reviewed reports from Oregon, Idaho, Kentucky, northern California, and interior British Columbia. Different terrain. Different populations. Same core descriptions. In fact, when we layer these descriptions onto maps and look for overlapping details, certain phrases cluster in surprising ways. The term "dead silence" before an encounter? Overwhelmingly concentrated in boreal regions. The description of "shoulders too wide for the head"? Common across Appalachian ridge sightings. The phrase "black hair with a reddish shimmer in the light"? That one repeats almost identically in Ontario, Washington, and the Caucasus Mountains of southern Russia.

This isn't just anecdote anymore. It's data. And if we treat the witnesses as field reporters instead of outliers, the consistencies speak volumes. The phenomenon, whatever it is, isn't just visible— it's *knowable*. Not by name. Not yet. But by shape. By motion. By behavior. And most of all, by the words people keep choosing—without even realizing it.

That's why I started keeping track of the phrases themselves. Not just what people saw, but how they said it. I've got notebooks where I've written these quotes line by line. Some from old letters. Some from email. Some from men too uncomfortable to speak, who just pointed and nodded while I read their own words back to them.

And here's what you begin to feel, after enough of that:

It's not that these people believe in Bigfoot.

It's that they don't *know* what else to call it.

The descriptions match because the thing itself doesn't change. Only our names for it do.

Chapter 3: Vocalizations That Shouldn't Exist
Part 1: Sounds Without a Source

The first time you hear it, you think it's someone playing a joke. A long, low howl—too deep for a coyote, too long for a wolf, too clean for a person. But then you hear it again. And it doesn't sound like anything. Not anything you've ever known. That's when your skin changes.

Audio is the one thing that leaves people rattled more than visuals. A distant figure can be misjudged. A footprint can be faked. But sound—raw, unfiltered, open-air sound—has a way of bypassing doubt and going straight into the body. The gut reacts before the brain has time to rationalize. And what witnesses describe over and over again isn't confusion. It's biological fear.

You hear it in their voice when they talk about it. They don't call it a scream. They say it "wasn't right." They say it was "too loud," "too long," "too low." It sounded like it came from something with a chest larger than anything human. A voice built on mass.

I've heard it myself.

Temagami, 2019. High ridge camp, just before 3 a.m. No wind, no rain. My audio recorder caught it before I did—three ascending whoops, then a pause. Then a rise into a howl that held for nine full seconds. Nine seconds. I've timed it a hundred times. It never clips. It just holds, then drops. No trail off, no panting, no follow-up. Just the kind of sound you don't expect to come from anything alive, let alone anything bipedal and hidden.

I've spent years comparing field recordings. Mine and others'. Ontario has produced dozens—some clearer than others, some questionable, many baffling. Spectrogram analysis shows frequency ranges overlapping with moose and large canids—but extending well beyond what they can sustain. The harmonics are wrong. The length of the vocal bursts defies any known animal in this biome. And perhaps most notably: the spacing between calls is too deliberate.

These aren't howls for communication. They're *statements*.

And they usually happen when no one's supposed to be listening.

Part 2: What the Recordings Reveal

When you isolate the audio, strip away the wind and insects, and look at it on a spectrogram, the story changes. You're no longer listening to a spooky sound in the woods. You're analyzing a signal. And that signal often defies classification.

One of the most famous examples came out of Snohomish County, Washington, in the 1970s—the so-called "Samurai Chatter." A series of guttural, rapid-fire vocal exchanges that sounded half-human, half-gibberish. Linguists couldn't place the language. Audio engineers couldn't find natural analogues. It was dismissed by many as too bizarre to be real. And yet, I've heard similar cadence in a recording from 2021 in the Kawarthas—so similar it made my stomach drop when I first heard it.

That Ontario clip was picked up by a camper with no interest in Bigfoot. He thought it was someone messing around in the bush, until he realized he was thirty kilometers from the nearest access road, and the "voice" was never followed by movement. No twigs. No brush. Just vocal bursts. Rough, guttural, rhythmic. Then silence. Then one more burst, much closer.

He left early.

I've run that recording through frequency filters, matched it against all known Ontario fauna. It doesn't match wolves, bears, elk, or moose. It doesn't match human voices either—at least, not in pitch or pattern. The vowel spacing, the breath support, the duration—everything suggests a lung capacity far beyond ours. And the spacing? Almost conversational.

Another example: a night call captured outside of Ignace, near a freshwater bog. A long, mournful cry, repeated twice, then followed by three rhythmic knocks spaced twelve seconds apart. The sound was recorded at 2:41 a.m., with ambient temps just above freezing. There were no other camps for miles. The call triggered the microphone's gain limiter. That means it was within 150 feet and *loud*. You can hear water moving in the background, wind passing over shallow reeds, and then—clear as breath—that cry. It's low, but not low enough to be a bull moose. And there's a tone to it. Not quite sorrow. Not quite alarm. But something… *aware*.

I've collected a half-dozen similar recordings across Northern Ontario, and they don't match each other perfectly. That's what makes them interesting. There's variation. That's what you'd expect from individuals—not from a looped animal call. Some are higher, some sharper, some cut off mid-phrase like they were interrupted. But they all share the same core traits: deep resonance, long breath length, and an eerie sense of modulation. Like a throat that knows what it's doing.

A wildlife biologist I work with reviewed one of these files. He said, "If that's not fake, it's coming from something bigger than anything we've catalogued in this forest." He doesn't believe in Bigfoot. He just believes in what his ears and equipment are telling him.

And that's where we are now.

Recordings are mounting. Independent ones. From credible people. Campers. Rangers. Hunters. Not Bigfoot hunters—just people who were out there and heard something they couldn't explain. The internet is flooded with garbage. But tucked inside that noise are real, vetted, unexplained pieces of audio that repeat over time, over geography, and across species barriers. They aren't always long. They're not always loud. But they share one trait that science can't quite dismiss:

They don't belong to anything we've catalogued. And they keep showing up exactly where the sightings do.

Part 3: When People I Trust Heard It Too

There's a moment in every conversation when someone shifts from skeptical to shaken. It usually happens after they've heard the sound themselves. Not on a recording. Not on a podcast. In the bush. In the dark. With no one else around.

You can hear the change in their voice when they talk about it. They stop trying to be scientific. They stop trying to explain. They just go quiet. And then they say the same thing I said after my first time:

"It wasn't anything I've ever heard before."

That silence—that hesitation—is more telling than any recording.

In 2020, I was out near Lake Wanapitei with a former conservation officer. Retired. No nonsense. He'd spent decades in the woods and could identify almost any bird or mammal by sound alone. We'd set up camp off a decommissioned fire road, just far enough off the main trail to lose the drone of civilization. No phones. No noise. Just a watch and our gear. Around 1:20 a.m., something let out a howl from across the valley. Long. Even. Not panicked. Not aggressive. Just… *big*.

He looked at me and didn't say a word.

Then it happened again. This time closer. We stood still, backs straight, like deer before a break in the treeline. He finally said, "Not moose." Then, a moment later, "Too round for a wolf. You hear that tail-off?"

I nodded. He just said, "Never heard that in my life."

He never called it Bigfoot. He still doesn't. But that moment stuck with him. And a few months later, he called me—just once—to ask if I had any recordings from earlier trips in that region. Not to prove anything. Just to *compare*.

That's how the shift begins. Not in belief—but in the unwillingness to dismiss.

Same thing happened in 2017 with two Algonquin guides north of Mattawa. They were camped near a river mouth and heard rhythmic knocking for close to ten minutes. No wind. No woodpeckers. No echo. Just consistent impact. One of the guides, a man who had spent over thirty years in the bush, said he felt it in his chest before he heard it in his ears. Said the knocks were so even they felt like code.

He asked me, weeks later, if there were any other cultures that used tree knocks for communication. He wasn't interested in YouTube videos or theories. He wanted anthropology. That's where his mind went—not monsters, not myths. *Language*.

Another contact, a paramedic from Atikokan, told me about a late-night call out on a rural road. Nothing Bigfoot-related. Routine motorist injury. But while loading the patient, they heard something howl from the forest line so loud it cut through the ambulance like a pressure wave. The patient asked if it was a siren. The medic told me he lied and said it was a moose.

Later, when I met him in person, he told me the part that didn't make it into the report: the injured man, in a daze from blood loss, mumbled something that stuck with him. He said, "Don't let it see me."

These aren't cryptid chasers. They're trained professionals. Field-experienced, level-headed people with no desire to be part of this narrative. And yet, here they are. Remembering sounds they can't forget. Reaching out quietly, privately, just to check if someone else heard it too.

They didn't need to be convinced.

They needed to know they weren't alone.

Part 4: What Could Make These Sounds?

At some point, you stop asking *if* the sounds are real and start asking *what could make them*. And that's where things get complicated. Because if you take the witness accounts seriously—and the recordings even more so—you're left with a shortlist of possibilities. None of them are comfortable.

Let's start with what we know. The forests of Northern Ontario are home to a wide array of vocal mammals. Wolves, moose, lynx, elk, bears, coyotes, and the occasional wandering mountain lion. Each one has a vocal signature. Some are more expressive than others, and each can produce terrifying, almost unnatural sounds when startled, threatened, or in heat.

But they have limits.

Moose calls, for example, have a nasal, almost hollow quality. They're powerful, yes—but they don't hold a tone for more than a few seconds, and they lack modulation. Wolves have beautiful, eerie howls, but their calls tend to have multiple voices in a pack and usually rise and fall quickly. Bears grunt, bellow, even roar—but they don't *howl*, and they certainly don't knock on trees.

And then there's breath control. That's the quiet tell.

Some of the recorded calls from Ontario and beyond last seven to ten seconds—long, controlled exhalations with consistent tone and volume. That implies a massive lung capacity, one well beyond what a human could sustain without training, and far past what a typical mammal in the region could manage. To create that kind of sound in one breath, the animal must be both massive and upright in its posture—letting the diaphragm expand in a way quadrupeds can't.

And what about the knocks?

No known animal in North America deliberately strikes wood with wood. Woodpeckers, sure—but their rhythm is erratic, echoing, high-pitched. These are low-frequency thuds. One knock. Then three. Then silence. Sometimes spaced a minute apart. Sometimes responding to a knock from another direction. I've personally documented these occurrences in Noganosh, Rabbit Lake, and northeast of Hearst. There are no matching species behaviors for this. None. Not even among primates native to other continents.

That leaves us with two choices.

Either an undiscovered animal is responsible—something large, intelligent, and reclusive—or humans are manufacturing every one of these encounters, recordings, and sound patterns over decades, across continents, with zero physical reward and incredible attention to detail.

The hoax theory falls apart under weight. Who's hoaxing calls in a swamp at 3 a.m. with no audience? Who's knocking on trees twenty kilometers from a road, with no idea someone might be there to hear it? Why would multiple strangers describe the same cadence, tone, and effect from independent encounters over generations?

They wouldn't. They couldn't. Not without a script.

And no one's handed them one.

That's why serious researchers, even skeptics, are paying more attention to sound. It's unfiltered. Raw. Hard to fake convincingly under scrutiny. And it often reveals one unavoidable truth: something out there is making sounds we cannot classify. And it does so in places where people see large, upright figures. In places where tracks are found. In places that go silent seconds before the sound arrives.

We may not have a name for the source yet.

But we've been hearing it for decades.

And now we're finally listening.

Chapter 4: Tracks, Prints, and the Weight of Evidence

Part 1: Impressions That Shouldn't Be There

Footprints are the oldest form of evidence we have. Long before trail cams, drones, and DNA swabs, there were just prints in the ground—and the uneasy realization that something too big, too heavy, and too human-shaped had passed through a place where nothing should've.

They still appear. Every year.

Sometimes it's just one or two prints—half impressions on soft trail edges, pressed into moss or loam, barely visible unless you know how to look. Other times, it's full tracks. Step lines. Twenty, thirty feet of stride over mud or snow or sand. Some are deep enough to show clear toes. Some reveal mid-tarsal flexibility, something modern humans don't typically show, but great apes do. Some are so large—14, 16, even 18 inches—that you'd need prosthetics or carved molds to fake them, and even then, you wouldn't get the depth. Because the depth is the key.

The deeper the print, the more pressure had to be applied. And in forest soil—especially damp muskeg or pine floor—you can't fake pressure. You can fake shape, but not displacement. Not the compaction of substrate under weight.

One of the best trackways I've seen came from a family visiting their cottage in December of 2013. They took a video of a trackway right on the road—fresh, clean impressions laid across a couple of inches of snow. The trackway stretched out in front of their vehicle, with approximately a hundred prints visible as they slowly drove along the cottage road. What made the footage so compelling wasn't just the number of prints—it was the family's commentary. Completely unscripted. You could hear the genuine confusion and awe in their voices as they tried to make sense of what they were seeing.

Eventually, they got out and measured the prints. Each one was 15 inches long and 8 inches wide. Wide enough to stand out immediately from any human track, and long enough that even a large man in boots couldn't have made them—especially not with that kind of uniform spacing. The gait was steady, the depth consistent. No sign of dragging, no doubling back, no stagger. Just a clean line of massive, barefoot impressions through untouched snow.

Incredible.

I've cast dozens of prints over the years. Some come out clean. Others fragment, the soil too loose or shallow to hold form. But the best ones—like the cast from Rabbit Lake in 2014, or the partial snow track line outside of Foleyet in 2019—hold details you can't explain away. A narrow heel. A thick midfoot. Toes that spread on impact. That's not a costume. That's an evolved walking system.

And here's what people don't realize: fake prints are usually easy to spot. They lack pressure points. They look too symmetrical. They show no skin folds or shifting weight. And most importantly, hoaxers don't account for terrain. They forget how real animals move through variable ground. The step spacing changes. The angle adjusts. Real prints *respond to the land*.

That's how you know.

That's how I knew.

I'm not saying every print is real. I'm saying the real ones don't go away. They don't fade from record. They accumulate. Quietly. From Ontario to Washington. From Alaska to the Carolinas. From places you'd never think to look—until you do. Then suddenly, there they are. Like a shadow with mass. A myth that leaves weight behind.

And once you see one—*really see one*—you stop asking *why people believe*.

You start asking *why no one's listening*.

Part 2: The Casts No One Talks About

In any other field, casts would be treated like evidence. Plaster molds pulled from deep wilderness terrain, showing detail down to dermal ridges and midfoot compression—that should mean something. But in the world of Bigfoot research, they're often brushed aside as curiosities, artifacts of a fringe hobby. Not because they lack quality—but because of what they imply.

And that's the real issue.

If even one of these casts is authentic—just *one*—it means there's something out there we've failed to categorize. Something bipedal. Large. Likely nocturnal. Avoidant. And intelligent enough to leave so few behind.

But the casts exist.

The earliest I personally examined was from 1984, near the Quebec border. Pulled from deep moss beside a trapline. Rough around the edges, but clearly showing five toes, a broad forefoot, and a strange break at the arch—a flexibility that didn't belong to any known North American mammal. At the time, it was dismissed. The trapper who took the cast was ridiculed. But that cast still sits in a storage locker just outside of North Bay, and when I held it in my hands, I saw something unmistakable: a print left by weight and motion, not by carving or mold.

In later years, I studied more technical examples. The famous "Cripple Foot" cast from Bossburg, Washington. The prints from the Blue Mountains in Oregon. And here in Ontario, at least six casts I've either made or examined personally—three of which I believe are legitimate. They all share common traits: pressure ridges, midtarsal break flexibility, and toe splay consistent with a heavy, unshod foot moving over natural terrain.

One cast, taken near Gowganda in 2017, changed the way I viewed this evidence entirely. It was made in shallow clay and sand, left behind by a receding creek. The toes were spread—not from static posing, but from motion. You could see the drag of the third and fourth digits as the foot pushed off the ball. A faint ridge of skin lines curved across the ball of the foot. This wasn't a static imprint. It was a step.

That's something very few fakes account for. Dynamic motion. The way real feet interact with real terrain. It shows up in the substrate, in the way the arch presses down, in how the toes flex unevenly as the foot rotates during a stride. Hoaxers usually stamp. Animals usually stomp. But these casts walk.

Many researchers—independent and academic—have begun collecting casts, either original or duplicated. Among the most well-known is Jeffrey Meldrum, Ph.D., a professor of anatomy and anthropology at Idaho State University. Dr. Meldrum is believed to have one of the largest cast collections in the world, with over 300 specimens. His archive includes casts from the Pacific Northwest, the American South, and parts of Canada, and he has conducted detailed anatomical analyses on their morphology. His work has brought legitimacy and academic scrutiny to an area long written off as pseudoscience. And like others in the field, he emphasizes the same key point: not all prints are equal—but some defy explanation.

And that tells us something else: the creature doesn't want to be found. These prints don't show up in clearings. They show up on forgotten trails, along quiet creekbeds, deep in the bush, or—like the 2013 snow trackway—by accident. When the animal slips up. Or maybe doesn't care for a moment.

That's what makes the rare casts so important.

They aren't symbols. They're biology. Clues. Like a hair snag or a scat sample—except clearer, more tangible, and deeply unnerving in their implications.

If this was any other field of study—zoology, anthropology, even primatology—these casts would be documented, preserved, reviewed. But because they come from *this* phenomenon, they're shelved. Mocked. Buried.

And still, they keep showing up.

They're not just impressions in dirt. They're the footprints of a mystery that walks like a man— but doesn't live like one.

Part 3: How the Prints Match Globally

It's easy to believe that this is a North American phenomenon. The term "Bigfoot" is culturally American. "Sasquatch" has roots in British Columbia. Ontario has its own local variations—"Old Yellow Top," "the Backwoodsman," "the Wildman"—but all of it seems tied to the continent. Until you start comparing the prints.

That's where the story gets wider. And older.

Because long before the internet, long before Patterson and Gimlin, other cultures were casting or sketching large, human-like footprints in remote regions. In the Himalayas, Nepalese sherpas spoke of the Yeti. In Siberia, the "Almasty." In China, "Yeren." In Australia, "Yowie." And when you remove the language, the mythology, and the regional detail, you're left with one unmistakable pattern:

The tracks match.

I've reviewed casts and documentation from multiple continents—directly and through credible research networks. In almost every case, the most compelling prints show the same anatomical features found in North American trackways: wide forefoot, absence of arch, flexible midfoot, and proportionally shorter toes that splay under pressure. The stride lengths vary slightly by region—likely due to substrate or slope—but the core mechanics remain consistent.

One cast from the Caucasus Mountains of southern Russia, taken in the 1980s, displays a footprint nearly identical in proportion to prints found near Mount St. Helens decades earlier. No shared media. No hoax coordination. Just mirrored physiology. A kind of convergent footprint language carved into the soil across continents.

In Australia, Yowie prints have been documented with similar depth-to-weight ratios as those found in Canada. Heavy, deep impressions in red clay and sandy loam, often in areas with no nearby human activity. Some even show signs of dermal ridges—patterns in the skin texture visible in a few of the best-preserved North American casts.

Skeptics dismiss this as coincidence. As if people from remote corners of the world just happened to fake the same type of foot, with the same pressure patterns, and the same mid-tarsal flexibility. But biology doesn't work in coincidence. It works in *function*. And the way these prints show up—with the same structural logic, in the same types of terrain, under the same conditions of stealth and avoidance—suggests function above all else.

In 2012, I had a conversation with an anthropologist from Finland who'd spent time with indigenous groups in Mongolia. They didn't have a name for what they described, but they spoke of a tall, hairy being that left long, flat footprints in the snow near the edges of tree lines. They showed him a drawing—five toes, slightly angled pinky, no arch. When he saw a copy of a Bigfoot cast from Idaho, he went pale. "That's it," he said. "That's what they drew."

It's not just prints. It's *matching prints*.

Separated by language. By culture. By oceans. But not by form.

So the question becomes: are we all inventing the same thing? Or are we all noticing the same thing, and just calling it by different names?

Because the prints don't lie.

And the further you follow them, the more it looks like this mystery doesn't belong to one region. It belongs to the world.

Part 4: Why No One Takes Prints Seriously Anymore

There was a time when a footprint in the woods could start a firestorm. Local news. Field investigations. Academic interest. That time has passed.

Today, a clear track in the snow or a plaster cast with dermal ridges barely makes a ripple. Not because the evidence is any less compelling—but because the noise around it has drowned out the signal.

We're in an era where everyone has a camera, but no one knows what to believe. Hoaxes are easier to create, distribute, and manipulate than ever before. And every fake footprint posted to social media—usually poorly made, symmetrical, and obviously staged—drags the real ones down with it. It's guilt by association. Even the authentic gets dismissed, because now the burden isn't just to be *convincing*—it's to be *undeniable*.

And in the world of Bigfoot, nothing is ever undeniable. That's the curse.

Add to that the fact that mainstream science has mostly turned its back. No academic career thrives in the shadow of Sasquatch. Even researchers with impeccable credentials—like Dr. Jeff Meldrum—are routinely ridiculed simply for cataloging and analyzing casts. Never mind that his work is anatomical, biomechanical, and peer-reviewed. The moment "Bigfoot" enters the conversation, the door to serious recognition slams shut.

It's easier for the public to mock than to consider the implications. Because if just one print is real, then something's been walking the forests of North America, undetected and undocumented, for decades. And that thought—that we might've missed something so big for so long—is terrifying to institutions that pride themselves on cataloging the world.

So they ignore it.

And the internet—fueled by short attention spans, clickbait, and crypto-entertainment—turns it all into spectacle. Someone finds a print? Upload it. Add dramatic music. Suggest aliens. Or cloaking. Or mindspeak. Turn a moment of field-based observation into a paranormal circus.

Meanwhile, the real casts sit in boxes. The plaster yellows. The records fade. And the people who actually saw the prints—the ones who crouched over them in disbelief, measured them with their hands, took photographs they never showed anyone—quietly walk away from the conversation. Not because they've stopped believing. But because they're tired of not being believed.

It's a shame.

Because prints are the most physical, objective evidence we have. They're not stories. They're not grainy videos. They're geometry. They're anatomy pressed into the earth. And they should be the foundation of real inquiry—not fodder for memes and message boards.

But until the noise dies down, until hoaxes fade and the serious voices rise again, the value of these prints will remain buried.

Even though the truth might still be sitting in a mold somewhere.

Even though the proof might already be under our feet.

Chapter 5: Cameras, Drones, and the Limits of Technology

Part 1: The Evasion Blueprint

Why they avoid cameras, roads, and open areas. Includes AI image recognition failure cases, scent avoidance, and behavioral mimicry.

If Bigfoot is real—and I believe it is—then the biggest question isn't why we haven't found it. It's why we haven't filmed it. Not clearly. Not conclusively. Not in a way that forces institutions to take notice. And that's where the story takes a darker, more frustrating turn. Because I've spent years trying to outsmart something that never seems to trip the wire.

They avoid cameras. Consistently. Game cams, thermal units, passive trail recorders—it doesn't seem to matter. Either they spot them and turn back, or they know how to move around them. And if that sounds far-fetched, consider this: deer and coyotes routinely detect trail cams. Bears have been known to approach and swat them. Animals notice changes in their environment—especially ones that carry human scent or infrared output.

Now imagine an animal with intelligence just past that of a chimpanzee. Something observant, slow-moving, cautious, and possibly aware that humans set traps. It wouldn't take long before it began to recognize which areas were "wired." And once that happened, we'd never catch it the same way twice.

I've placed dozens of trail cams in northern Ontario. I've camouflaged them, masked the scent, run them on thermal-only triggers, even buried them in brush and aimed them toward bait zones. Over the years, I've captured moose, bears, wolves, hikers, illegal loggers, and once—a rare white moose moving silently through fresh snow at dusk, its coat glowing like a ghost in the camera's IR flare. But never anything resembling a Bigfoot. And I've set these cams in areas where sightings were reported within 48 hours—same weather, same terrain. Still nothing.

The failure isn't always technical. Sometimes, it's behavioral. They don't walk down trails. They don't move in the open. They cross roads only when the timing is perfect—late, silent, and fast. Multiple roadside sighting reports note the same detail: the figure stepped out, crossed in two or three strides, and was gone. Never a pause. Never hesitation.

That kind of movement doesn't just suggest intelligence. It suggests *an understanding of how to avoid detection.*

Some researchers believe this is learned behavior—adapted over generations from close calls with humans. Others think it's instinctive. Whatever the cause, it creates a profile of an animal that is not just elusive, but actively evasive. It doesn't stay in one spot. It doesn't linger near populated areas. And it doesn't tolerate the artificial. That includes drones.

Now, drones have become a hot tool in recent years. They offer a bird's eye view, thermal imaging, and extended tracking potential. But again, they've produced almost nothing. Why? Partially, it's because thermal imaging struggles through canopy. And partially, because we don't know where to look. But there's also this: the sound of a drone carries. It buzzes. It pulses. And in the backwoods of Algoma or the boreal ridges of Temagami, that sound travels like a siren.

And here's where it gets more interesting.

Several reports—including my own field experiences—suggest scent avoidance plays a major role. You can be quiet. You can move slowly. You can mask your lights. But you can't mask your smell—not fully. Multiple witnesses have described Bigfoot encounters ending the moment they lit a cigarette, unzipped a backpack, or cracked open a protein bar. The figure turned and walked away. No chase. No curiosity. Just immediate withdrawal.

And then there's behavioral mimicry.

A handful of reliable reports suggest that these creatures observe humans from a distance. Long enough to pick up on patterns. Not just movement, but rhythm. One story from 2015 near Hearst described what sounded like a person chopping wood with a heavy axe—regular, spaced thuds. Except there was no one there. When the sound was followed into the trees, it stopped. No tracks. No scent. Just silence. If that was a creature mimicking behavior to mask its own movement, we've crossed a threshold: not just evasion, but deception.

And AI? Facial recognition? It's not ready. I've fed dozens of trail cam images—blurred, partial, backlit—into commercial-grade recognition software. AI fails every time. Why? Because there's nothing in the database that looks like this. It wants to assign a known category: bear, human, moose. If it can't, it flags it as an error or crops it out as noise.

That's not just a limitation. That's a blind spot.

Which means the most advanced tools we have in 2025 still can't find what a frightened hunter saw with his bare eyes fifty years ago. Not because the technology is flawed—but because the subject refuses to cooperate.

It doesn't want to be filmed.

It doesn't want to be found.

And somehow, it knows how not to be.

Part 2: Modern Surveillance Fails—What the Tech Misses

We've convinced ourselves that technology can solve anything. Satellites, drones, AI, motion sensors—if something's out there, we'll catch it. That's the promise. And it's a good one. But it hasn't held up.

In practice, surveillance technology fails more often than we admit. Not because it's broken, but because it's built for the wrong target. It's designed to detect *known* things—human faces, thermal bodies, familiar shapes. But Bigfoot, assuming it's real, isn't cooperating with those expectations. And neither is the environment.

Take drones. With high-resolution cameras and thermal optics, they should be perfect for scanning vast forested areas. But most drones can't see through canopy. They give you a perfect view of treetops—green, endless, and useless. Heat signatures don't penetrate down through the layers unless there's an open break in the foliage. And even when there is, most animals hunker low, stay still, and disappear into the background noise. By the time the drone circles back for a second pass, whatever was there is long gone.

Motion sensors? Better in theory than in practice. In the dense Canadian bush, everything moves. Wind, squirrels, insects, low-flying birds. Cameras trigger constantly, burning battery life and filling cards with noise. I've reviewed thousands of these images—blurred wings, swaying branches, raccoons, porcupines, the occasional fox. The one time something *almost* looked strange—a shadowy bipedal shape partially blocked by a birch tree—it was already out of frame by the time the second image fired. One blurry shoulder. One foot in motion. Gone.

Thermal scanners? More reliable in open terrain, but in the forest, you're dealing with a chaotic temperature map. Rocks retain heat, water reflects infrared, and trees—especially in spring and fall—create temperature gradients that make tracking nearly impossible unless your subject is moving fast and unobstructed. And Bigfoot, by most reports, doesn't do fast. It does quiet. It does slow. It stays in shadow.

Even satellite imaging, with all its promise, breaks down at the practical level. The resolution isn't good enough to pick up individuals under foliage. And the data isn't real-time. By the time you analyze what looks like a heat anomaly, it's hours old—and likely just a moose lying in a sunlit clearing.

Then there's AI. Pattern recognition software has come a long way. It can find missing hikers from above. It can track wolf packs by coat variation. It can ID license plates from orbit. But it chokes on ambiguity. Show it a blurry, upright figure in a forest shadow, and it hesitates. It either classifies it as "human" with low confidence, or discards it as an environmental artifact. In testing, I've fed AI over a dozen allegedly credible Bigfoot photos—most returned "low-certainty human" or "environmental noise." One flagged as "possibly bear." None saw what the human eye saw instantly: a vertical figure standing out against the geometry of the woods.

Tech fails in subtler ways, too.

Batteries die faster in cold. Sensors fog. Cameras freeze. SD cards corrupt. Even if you *do* capture something unusual, you then have to *extract* it. Which means going back into the bush, sometimes alone, hoping the unit is still there, intact, and undisturbed. I've lost two entire data sets that way—one stolen by who-knows-what, the other crushed by a falling pine.

And this is where belief collides with frustration.

If we were looking for *any other* rare species, we'd acknowledge these gaps. We'd factor them into the margins. But because we're looking for Bigfoot, the standard isn't just high—it's impossible. "Why don't you have a photo?" becomes the end of the conversation, instead of the start of a serious inquiry into why traditional methods are failing.

The truth is, modern surveillance wasn't designed for an evasive, intelligent animal that has adapted to avoid *us*. It was built to monitor, to confirm, to classify. Not to track something that's trying not to be known.

Until we change how we look—how we search, how we filter, how we interpret—we'll keep missing what might be just behind the tree line. Watching us. Waiting for the drone to pass.

Part 3: The High-Tech Arms Race in the Woods

The deeper into this subject you go, the more it begins to feel like a cold war—researchers on one side, pushing farther and harder with new tech, and something else on the other side, quietly outpacing every advance.

That's not hyperbole. It's field reality.

Trail cams got smaller? Whatever we're tracking started avoiding them. Thermals got sharper? The sightings moved to thicker canopy and colder months. Drones got quieter? The figures stopped appearing in open spaces. With every leap in equipment, the distance closes—and then the subject recedes again, just out of reach. Like it knows.

This has pushed some researchers into what I call the "tech spiral"—a cycle of constantly upgrading, constantly investing, chasing that perfect setup that might finally break the mystery. And make no mistake, the tools are impressive. I've worked alongside teams running full-spectrum video rigs, aerial thermals, and autonomous listening stations capable of logging sound profiles over weeks. I've deployed AI-assisted audio filters that pick up unusual vocal patterns— long whoops, repeated knocks, guttural phrases in subsonic ranges.

And yet the results are always the same: **close, but no confirmation**.

Take acoustic surveillance. In 2021, a colleague of mine set up a remote, solar-powered audio array near Lake Abitibi. It ran 24/7 for three weeks, capturing nearly 900 hours of forest audio. Dozens of vocal anomalies were logged—some matched known animals, others didn't. One stood out: a double knock followed by a low, single whoop and a long silence. It didn't match any known species. The waveform resembled some calls collected in the Sierra Nevada years earlier. We isolated it, triangulated its position, and returned the next night with three high-resolution cams set in a triangle.

The result? Nothing. No return vocalizations. No heat signatures. The woods went dead quiet.

That kind of retreat isn't random. It's reaction.

Some teams have experimented with passive LiDAR mapping—scanning forest environments for structural movement or changes in topography consistent with large bipedal movement. But the data sets are massive, and anomalies still rely on human interpretation. You might catch a strange, upright shape in a scan, but you'll never prove what it was unless you're there on the ground—something this subject seems determined to prevent.

Even private funding has begun shifting toward custom UAVs: long-range drones with onboard AI filtering and real-time data uplink. The hope is that faster data will mean faster decisions—that if something strange is detected, you can move in before it disappears. But here's the problem: time and again, the actual sightings happen in places *just outside* our grid. Not by accident. By *design*.

And we don't talk enough about what that implies.

If this thing is simply an animal—flesh and blood, as I believe—then it is an animal that has somehow learned, or evolved, to avoid human detection systems. Not just people. Our *tools*. Our *patterns*. That's not paranormal. That's *strategy*.

So we upgrade. Every year. We chase with better optics, better algorithms, quieter drones, smarter cams. And somehow, it's always *one step ahead*.

The irony is that in trying to corner it with tech, we may have only taught it what to avoid. We've triggered the arms race. And now, we're not just tracking something mysterious.

We're chasing a ghost with a playbook.

Part 4: When Cameras Fail—What Eyewitnesses Still Get Right

We've put our faith in gear—thermal, trail cams, drones, AI. But when the batteries die, the sensors freeze, or the lens captures nothing but trees, it's still the human witness who holds the line.

And the truth is, they're getting more right than we give them credit for.

There's this idea floating around—that eyewitnesses are unreliable. That memory is flawed. That adrenaline distorts detail. And sure, sometimes it does. But in the hundreds of interviews I've done, from seasoned trappers to backcountry campers to reluctant locals who just wanted to get it off their chest, a strange consistency emerges.

They don't embellish. They hesitate. They choose their words carefully. They pause too long before they speak.

And when they do speak, they often describe the *same things*.

The same gait—long stride, knees slightly bent, arms swinging lower than a human's.

The same body language—straight-backed posture, quick but deliberate movement, no side-glancing or flinching like an animal surprised.

The same reaction—*it saw me, and then it left*.

These details don't come from internet forums. Many of these witnesses had never read a Bigfoot book. Some didn't even know what they saw until someone told them after. And yet they report the same posture. The same eyeshine height. The same silence in the woods right before the encounter.

What they describe holds up *because* it's so specific. Too specific to be fantasy. Too strange to be random.

I spoke with a retired MNR employee in 2016 who had an encounter east of Smooth Rock Falls. He was surveying beaver damage when he crested a ridge and spotted a figure standing by a blown-down pine. He froze. No camera. No binoculars. Just his eyes and a clipboard. He described the shape in fragments: wide shoulders, no neck, cone-shaped head, arms longer than they should've been. He said it stood like it didn't care. Not afraid. Not aggressive. Just... *watching*.

And then it turned. Took three steps. Gone.

A camera would've failed there. The shot would've been obscured. Blurred. Or it wouldn't have been ready at all. But his description, weeks later, matched others I'd collected from the same region. Almost word for word.

And that's what people don't understand: when enough people see the same thing, in different places, at different times, without ever having compared notes—that *means* something. That's not imagination. That's data.

Eyewitnesses still get details that elude our sensors. They describe behavior. Movement. Emotion. How the thing reacted to them. How it tilted its head. How it froze at the edge of light, almost calculating.

Machines can measure. But people can feel. They notice things machines miss—like the smell that hangs in the air after, or the dead quiet before it appeared, or the way their dog, loyal and fearless, refused to move an inch when it stepped into view.

There's a reason cameras haven't replaced testimony.

Because whatever's out there still knows how to avoid the lens.

But it hasn't yet figured out how to avoid *being seen*.

Chapter 6: Hair, DNA, and the Frustration of the Lab

Part 1: The Samples That Lead Nowhere

If footprints are the body, and eyewitnesses are the voice, then hair and DNA should be the fingerprints—the final proof. The clincher. The biological anchor that connects sightings and stories to science.

But that's not how it's gone.

Because in the search for Bigfoot, DNA doesn't clarify. It confuses. It fails. It vanishes. And the more promising the sample, the more frustrating the result.

Let's start with the hair.

Over the years, dozens of hair samples have been collected from alleged encounter sites. Fence snags. Branch rubs. Bedding sites. Even bark scratches. I've personally pulled what looked like thick, dark brown or black strands—longer than human, coarse, slightly curled—from a snapped alder branch near the Montreal River in 2018. No scat. No prints. Just a single hair caught in the bark at shoulder height. I took it home, bagged it, and sent it off for microscopy.

The result? "Unknown primate characteristics, but inconclusive." No root. No nuclear DNA.

And that's the pattern.

Over and over, labs report similar outcomes: *Hair with primate-like features, medulla structure inconsistent with known species, but no follicle. No usable DNA.*

That's not failure of the witness. It's the limits of forensic technology. Most commercial labs are built to identify known species. They use reference databases. If your sample doesn't match bear, wolf, human, moose, dog, or anything else in the system, it doesn't come back as "new species." It comes back as *no match*.

And "no match" means nothing to a skeptic. But to a field researcher, it's maddening. You've got physical material—tissue. And the machines say: *we don't know what this is, so it's nothing.*

Scat is even trickier. First, it's rare. Second, it's almost always contaminated. By the time it's found, insects have touched it. Rain has soaked it. UV light has degraded it. And if you do get

DNA from it, it often shows what the animal *ate*, not what it *was*. A moose hair. A rabbit bone. Bark. That doesn't help you build a profile of the creature that left it behind.

Then there's the "mitochondrial mismatch" problem.

In a few rare cases, researchers have sent samples that returned *partial* sequences. The mitochondrial DNA—passed through the maternal line—sometimes shows human. But the nuclear DNA, when present, comes back fragmented, unreadable, or non-human. That leads to two possibilities: contamination, or something we don't have a reference for.

Mainstream science says contamination. And sometimes they're right. But not always.

Because here's what doesn't get said often enough: the lab doesn't know what it doesn't know.

If you send them a sample that doesn't match any of their cataloged organisms, they don't tell you it's Bigfoot. They tell you it's *failed*.

Which leaves us in the same place, year after year.

With strange hairs that don't match bear or man.

With tissue that turns up "primate-like" but inconclusive.

With samples that *should* be everything—but end up nothing.

The promise of the lab is that it will deliver truth. But in this field, the lab has delivered mostly silence.

And sometimes, that silence feels deliberate.

Part 2: Lab Gatekeeping and the Reluctance to Acknowledge the Unknown

In theory, science is open-ended. It's supposed to follow the evidence wherever it leads. But in practice, the moment you send in a sample with the word "Bigfoot" attached, the door starts to close—quietly, and fast.

I've seen it happen.

In 2019, I worked with a small independent lab out of Toronto. We submitted a hair sample pulled from a bark scrape in Northeastern Ontario. The field collection was clean—gloves, tweezers, sealed vial, labeled coordinates. When we submitted the paperwork, we used only the GPS info and said it was from "an unidentified animal of interest." The initial response was professional. Curious, even. But the moment they started asking questions—location, purpose, origin—and we told them we were investigating a large, upright animal reported in the area, their tone changed.

They stopped replying to follow-ups.

No official report was ever delivered.

When we called, the receptionist said the sample had been "destroyed in processing."

Now, maybe that's all it was—bad luck, bad handling. But it wasn't the first time.

This pattern repeats across North America. Good samples vanish. Labs withdraw. Results go "inconclusive" or "unusable" without explanation. And when you press, even politely, the answer isn't science—it's attitude.

Because the moment Bigfoot enters the room, credibility leaves it.

And that's gatekeeping.

It's not a conspiracy. It's worse than that. It's culture. It's reputation. Labs don't want their names attached to fringe topics. Geneticists don't want to publish on something that could be seen as pseudoscience. So instead of engaging with the evidence, they shut down the inquiry.

In one case, a respected university returned a report on a fecal sample from the Pacific Northwest with a curt note: *"We suggest this be resubmitted under a more scientifically grounded context."*

That's not peer review. That's prejudice.

The irony is, these same institutions push the limits of zoology all the time. New frog species in the Amazon. Deep-sea life pulled from hydrothermal vents. Tiny primates hiding in Madagascar. That's celebrated. That's discovery.

But a 7-foot hominoid in Ontario?

That's crazy. That's career suicide.

Even labs that specialize in cryptid research have started distancing themselves. The moment a sample yields ambiguous results—partial human DNA, unidentifiable protein chains, mitochondrial anomalies—they're labeled contaminated or dismissed as "unresolved." No follow-up. No invitation to repeat the process under controlled conditions.

And no curiosity.

That's the part that stings most. Real science should be curious. When something doesn't match known taxonomy, the response should be: *why not?* Instead, we get: *must be error.*

I've spoken with biologists who are privately fascinated. They'll whisper their own sightings, share stories from fieldwork, admit that some reports make sense. But publicly? Nothing. Because publishing on Bigfoot could end their grant cycle. It could stain their career.

So they look away. They file the mystery under folklore. Or worse—ignore the data entirely.

That's how we lose progress. Not through a lack of evidence, but through the refusal to face it.

Gatekeeping in the lab isn't just about control. It's about fear. Fear of being wrong. Fear of ridicule. Fear of opening a door they might not be able to close.

Because if even one sample is real—just one—it changes everything.

And most institutions aren't ready for that.

Part 3: What a Breakthrough Would Actually Look Like

Forget what you've seen in movies. There won't be a dramatic press conference. No blurred photo held up under bright lights. No whisper of "we've finally found it." If a true biological breakthrough happens in this field—if we finally get undeniable proof of Bigfoot—it will start the way all real discoveries do.

Quiet. Boring. Technical.

It'll begin in a lab, under a microscope, or inside a sequencer. A hair, a scat smear, maybe blood. Something small. Something overlooked. It will be collected carefully—gloved, labeled, cold-stored. And this time, it won't be flagged as Bigfoot. It won't be marked with any buzzword that triggers dismissal. It'll go in as "unknown sample from boreal zone." No red flags. No eyebrows raised.

Then the processing begins.

The DNA will come back—first partial, then whole. The mitochondrial DNA might suggest primate. That's when a technician pauses. Looks again. The nuclear DNA might contain sequences never seen before. No matches in the human genome. No bear. No moose. No chimp. Just long strings of code that scream *new*.

A call gets made.

Maybe it goes to an independent lab, to verify. The second lab confirms the anomaly. No match. Reproducible results. Genomic patterns consistent with a higher primate, but split from the known family tree tens of thousands of years ago.

And here's where it diverges.

One path leads to silence—an internal memo, a shelved result, a sample labeled "archival." Nothing moves. Nothing changes.

The other path leads to publication.

Quiet at first. A genetics journal. A short paper. Unassuming title: *"Unidentified Primate DNA Profile from Northern Ontario."* Nobody panics. Nobody cheers. It's just a blip.

Then other researchers read it. Field biologists. Anthropologists. Someone notices the data doesn't make sense unless there's a living source.

Then the dominos fall.

Follow-up studies begin. The original sample is re-examined. Pressure mounts for location disclosure. The media gets wind. Headlines twist the science into clickbait—*"Bigfoot DNA Confirmed?"*—and suddenly the public is involved. Social media erupts. Skeptics and believers clash. The researchers retreat to protect their names. But the data holds.

And now it's out there.

At that point, cameras start pointing in the right direction. Not wild guesses, but focused surveys. Air support. Ground teams. Not weekend hobbyists, but funded expeditions. Academia begins to sniff around—still cautious, still smirking—but interested.

Then comes the second breakthrough: clear, repeatable footage. Maybe thermal. Maybe daylight. Maybe drone. This time, it's not fuzzy. It's clean. Long. With movement, behavior, gait. Anatomical consistency with the DNA profile.

It doesn't prove everything. But it proves enough.

And after that? The shift.

Textbooks get amended. A new species code is issued. Not Homo sapiens. Not Pan troglodytes. Something in-between. Or beyond. A hominin relative. A side-branch. A survivor.

Governments respond. Conservationists raise alarms. The debate begins about how to protect the creature—and how to admit we ignored it for so long.

But that's all later.

The real breakthrough won't be dramatic.

It'll be one sample. One sequence. One quiet realization in a quiet lab, that something unknown walked across our path, and left part of itself behind.

We just had to stop labeling it "impossible" long enough to notice.

Chapter 7: The Vocal Evidence—Strange Sounds, Calls, and What They Might Mean

Part 1: What We're Hearing in the Woods

Long before trail cams and DNA kits, before plaster casts and thermal drones, people were hearing things in the woods—things that didn't fit.

Strange howls. Deep, rolling whoops. Rhythmic wood knocks in dead still forests. Guttural moans echoing through valleys long after the sun had set.

These sounds didn't come with explanations. They came with silence. With stopped conversations. With the sudden stillness of birds and the ears-back retreat of dogs.

And the people who heard them? Trappers. Rangers. Loggers. Campers. Hunters. People who *knew* the woods. People who'd lived their whole lives listening to wolves, loons, coyotes, moose, owls. They weren't spooked by a cracked twig or a bear grunt. But these sounds—they rattled them.

Because these weren't animal calls. And they weren't human, either.

They were something else.

I first heard one in 2015, near a remote bog northwest of Foleyet. I was alone, hiking in to check a blind I'd set the week before. No wind. No birds. Then—clear as glass—a single, rising whoop from across the water. Deep-chested. Not mechanical. Organic. Long enough to feel, not just hear. I froze. My skin pulled tight. It didn't repeat. And that made it worse. It was like the forest exhaled something it shouldn't have.

That sound stayed with me.

I've since collected dozens of similar reports. Some accompanied by knocks. Some by movement. A few followed by sightings. But the common thread is the sound itself—unclassified, unfamiliar, unforgettable.

In the last decade, audio analysis has advanced. Field researchers have begun using long-range parabolic mics, night-recording stations, and real-time spectrum analyzers. And in that time, a few compelling clips have surfaced. Calls with frequency ranges far below human. Howls with harmonics that resemble great ape vocalizations—resonant, layered, impossible to fake with the human throat.

One series of whoops recorded in southern Washington was analyzed by a wildlife audio engineer in 2020. The waveform showed pitch shifts consistent with large-lunged mammals—elk or moose size—but the tonal structure didn't match *any* known animal in the database. The closest analog, strangely, was a chimpanzee long call—but slower, deeper, and spaced apart like deliberate speech.

And that's the unnerving part.

Some of these sounds seem patterned. Timed. As if they're communicating.

Not language, exactly—but intent. A challenge. A signal. A response.

Field teams have done controlled knock tests—producing loud wood-on-wood cracks and recording what comes back. In many cases, the return knocks are spaced to match. Two knocks, then two back. One, then one. Sometimes even a triple rhythm repeated perfectly. This isn't echo. It isn't coincidence. Something out there is hearing and answering.

In 2018, a camping group east of Chapleau captured a long, wavering howl followed by two knocks. The audio is unrefined—cell phone mic, open air—but when cleaned up, the howl registers just under 100 Hz at its lowest peak. That's *lower than most human males can vocalize.* And the knock that followed had the acoustic profile of a solid impact—like a baseball bat on a tree trunk, not a branch snapping.

Still, the scientific world won't touch it. Because sound is subjective. It can't be touched. Can't be weighed or bagged. It floats in and out, unprovable. And yet—it's one of the most consistent elements in reports across North America.

People are *hearing* something. And they're hearing the same things.

From Washington to Ontario. From Texas to northern Alberta. From Maine to Michigan.

And those who've heard it don't forget.

Because once you hear something scream from a tree line that's supposed to be empty, something that doesn't sound like any bird, mammal, or man you've ever known—you don't un-hear it.

You just listen differently after that.

Part 2: Sierra Sounds, Vocal Mimicry, and the Theory of Intentional Communication

The Sierra Sounds recordings were made in the 1970s by Al Berry and Ron Morehead in the remote high country of California's Sierra Nevada Mountains. What they captured has never been matched—dozens of minutes of rhythmic, vocal sequences echoing in the dark between trees. Some sounded like gibberish. Others came through as barked commands. There was breath. There was emotion. There was structure.

To this day, no credible source has replicated the audio with a human voice under the same outdoor conditions. Even skeptics admit: the recordings are real. The context is real. What's debated is the source.

Scott Nelson, a retired U.S. Navy cryptologic linguist trained in speech pattern recognition, spent hundreds of hours analyzing the Sierra Sounds. He concluded that the recordings exhibited qualities of an unrecognized language—syntax, repetition, phonetic variation. Not random yelling. Structured exchange. In his analysis, he suggested the calls weren't just noise—they were conversational.

This opened a question almost no one in mainstream biology wanted to touch: Could these animals, if real, possess the ability to mimic, or even generate, language?

That's where the discomfort begins.

Primates, including chimpanzees and orangutans, are capable of mimicry and basic sign systems. Some birds are master imitators. But a wild, undocumented primate issuing complex vocal signals in a cold forest at night? That demands not just intelligence—but a *social system*, one capable of developing shared sound structure over time.

The mimicry angle complicates things further.

Witnesses across North America report hearing their own voices mimicked back to them. Or campfire chatter replayed, slightly wrong. Dogs barked at empty woods. One Ontario hunter described a soft "hey" coming from behind a spruce wall—spoken in his missing brother's voice. But the brother had been dead six years. No one was there.

Others report more subtle vocal mimicry—birds that don't sound quite right. Coyotes calling in chorus, then finishing with a loud, low grunt no coyote could produce. A whistle that shifts tone mid-blow.

I've experienced it, too.

In 2020, northeast of Smooth Rock Falls, I was setting up an acoustic recorder along a swamp ridge. Alone. No animals visible. I let out two sharp whistles—just to test reverb. Five seconds later, two nearly identical whistles came back. Not echo. Not playback. Same spacing. Same tone. Slightly deeper, as if mimicked. Then silence.

That night, something walked close enough to my camp to leave two deep impressions in moss, right outside the tarp line.

This isn't unique.

In a 2023 study by independent researcher K. Monaghan, over 40 witnesses reported some form of vocal or sound mimicry within seconds or minutes of speaking, whistling, or knocking. These weren't hallucinations or faulty memories. Some were recorded. Some had multiple witnesses. In at least a dozen, the sounds were captured on audio but the mimic source was never located.

No scientist wants to say: this sounds like language.

But that's what it feels like.

Not speech. Not human. But intention. Directed sound. Reactive sound. Calls made with purpose. Responses given to questions that weren't even questions.

And the scariest part is how often it's *close*. Not miles away. Not across a valley. But right behind you. Just inside the tree line. As if it's not trying to lure you deeper, but let you know: *I'm already here*.

Part 3: Subsonic Speech and the Edge of What We Hear

Some of the strangest sounds reported in the field are the ones people don't realize they've heard—at least, not right away.

These aren't howls or whoops or obvious animal calls. They're pressure shifts. Deep pulses. Vibrations you feel before you process them as sound. Some researchers call them infrasound. Others call them subsonic vocalizations. Whatever the name, they occupy a frequency range **below** what the average human ear is built to detect—generally under 20 Hz.

We don't normally hear it. But we can feel it.

And in certain environments, it can induce fear. Anxiety. Nausea. Hallucinations. Even a sense of being watched or stalked. Not because of the situation—but because of the **frequency.**

Infrasound isn't new. Elephants use it to communicate over miles. Tigers can project it during roars—temporarily paralyzing prey. Whales use it in the deep. And military studies have long shown that subsonic pulses can affect the human body in profound, unsettling ways.

Now imagine something in the forest using it—deliberately.

There are multiple reports from hunters and campers in North America—Ontario included—who describe a "drop in the stomach," a "vibrating chest," or "feeling dizzy and scared without reason" *right before* a vocalization or movement event.

Not the moment after. Before.

In 2022, west of Hearst, a father and son on a bear hunt described hearing a low rumble, like distant thunder, even though the sky was clear. No planes. No trucks. Then, a single tree crack. Something moved behind their blind. The father raised his rifle, but his hands were shaking. Not cold. Just shaking. The son said it felt like "his ribs were buzzing."

Nothing came out of the woods. But when they checked the thermal cam later, it showed a single heat signature—tall, narrow, and still—moving away, then vanishing.

Another report from southern Ohio describes a family who were packing up their gear when the mother suddenly burst into tears—no trigger, no sound, just a wave of emotion and what she called "a drop in gravity." Ten seconds later, they heard three fast knocks and a deep, stuttering moan. They left immediately. The emotion was gone within minutes.

This is difficult territory for researchers. Because you can't see infrasound. You can't easily record it with consumer-grade field gear. And even if you *do* capture it, the proof is physiological, not visual.

But it's real.

In 2023, I tested an infrasound detection app on my phone—marketed as a utility for hunters and paranormal investigators. I didn't expect much. Phone microphones aren't designed to pick up frequencies below 20 Hz. But the app was built to *interpret environmental pressure changes* and visualize sub-bass shifts over time. I figured if anything anomalous happened, I might at least see a pattern.

I ran it on a late-season solo hike near Shining Tree, just before freeze-up. It was dead quiet. No wind, no birds, no gear running. Just forest. I left the phone on record mode in a pack, placed twenty feet off a well-used game trail. Came back hours later and scrubbed through the data.

Around 2:38 AM, the app logged an unusual low-frequency rise—starting at 18 Hz, peaking around 17.2 Hz, then flatlining back to baseline after about 15 seconds. Immediately after, the decibel meter spiked at a higher frequency: a *single, high dB knock*, exactly as loud and sharp as a wood-on-wood strike.

The log didn't explain what caused the infrasound. No aircraft, no storms, no recorded wildlife movement. But the timing was tight. It matched similar reports I'd heard from other researchers.

I've since used the app on six more outings. Only one other night showed the same signature. Always remote. Always after midnight. Always followed by motion in the tree line.

No one believes this is proof. It isn't. But it's evidence. It's a pattern. And it might explain why so many people freeze up during encounters. Why they feel sick or dizzy. Why their dogs tuck tail and run. Why they "felt something watching them" before they even heard a sound.

Because maybe they didn't hear it first.

Maybe they **felt it**.

And maybe that's intentional.

Not a roar. Not a howl. But a signal.

I see you.

Chapter 8: What People Say—"It Looked at Me Like a Man"

Part 1: Repeating Language from Witnesses Across the Continent

There's a phrase I've heard again and again from people who've come forward with encounters, whether from British Columbia or West Virginia, Michigan or Northern Ontario. Different people, different stories—but the line repeats like a drumbeat:
"It looked at me like a man."

They don't say gorilla. They don't say monster. They say man.

Sometimes it comes out through gritted teeth. Sometimes quietly, like they're confessing something. These aren't always people who believe in Bigfoot—or even want to. They're often the ones who *really* don't. And that's what makes the phrasing so eerie. Because they're not grasping for drama. They're grasping for something they can't explain.

It's not just the appearance they're talking about. It's the **look**—the way the creature turned, made eye contact, and held it. Not an animal glance. Not a reflexive flick like a deer or bear. But a direct, deliberate *look*—like it was assessing them. Like it knew what they were.

I've collected over 40 interviews in the last decade that contain variations of this exact phrase, most of them unsolicited. "Like a man." "Too human in the eyes." "I felt like it knew I was afraid." In some cases, the moment of eye contact was the scariest part of the encounter. Not the size. Not the sound. Not the movement. The **look**.

One man near Thunder Bay, Ontario, told me the creature stared at him from across a logging cut. Just stood there. Not hiding. Not running. And what unnerved him most wasn't the height or the shoulders—it was the stillness in the face. "It looked like a person who wasn't supposed to be there," he said. "But it wasn't a person."

I heard of a woman from Arkansas who used almost the same phrasing. She saw it crouched near a creek bed at dawn. When it turned, its eyes locked on hers, and she froze. "It wasn't animal eyes. It was like it was trying to decide something." She wept during the interview—not from fear, she said that part had passed—but from not knowing how to explain what she'd seen.

Some skeptics argue that this is just projection—people trying to humanize what scared them. But I've sat across from enough witnesses now to recognize real emotional dissonance. They're not trying to turn the creature into a person. They're struggling with the opposite: **how something that didn't look exactly human still managed to feel like one.**

They talk about its eyes. Not glowing. Not supernatural. Just aware. Deep set. Watching. Not scanning the terrain for food or threat—but *watching them*, with purpose.

This is also where you see splits in reported behavior. Black bears don't hold eye contact. They'll bluff, posture, even approach—but they won't look into you. Moose don't watch. They assess, then act. But these creatures? Witnesses describe the look as intelligent. Not curious—*aware*.

In 2019, I interviewed a retired conservation officer who said, without hesitation: "That thing looked at me like it was trying to figure out what I would do next. I wasn't the one stalking." He later refused to go back to that stretch of bush, even though it was part of his job for 30 years. Not because he was afraid of being attacked. But because of that look.

This part of the phenomenon—the expression, the stillness, the impression of consciousness—rarely makes it into scientific debate. Because it's subjective. You can't analyze a stare. But it's where belief is often born.

Witnesses don't need proof after that. Because in that moment, **they're the ones being observed**.

And that's not something they forget.

Part 2: The Anatomy of a Stare – Why the Eyes Break People

It's hard to describe the emotional damage a single moment of eye contact can do—especially when that eye contact comes from something you don't understand. Something you're not supposed to be seeing. Something looking *back*.

When I interview witnesses, I don't start with questions about hair or height or movement. I start with the moment they made eye contact. That's where the truth is. That's where the nerves unravel. That's the part people never forget. They might doubt the tracks later. They might explain away the smell. But the **look** stays. It burrows in.

And there's a pattern to how they describe it.

They don't say rage. They don't say hunger. They say **judgment**. They say it felt like the creature was reading them. Deciding whether or not they were a threat. Choosing what to do next. Not based on instinct, but on decision-making.

There's a biological reason this lands so hard.

In predators—bears, wolves, cougars—eye contact often signals a warning. It's quick, it's sharp, and it rarely lasts more than a second or two. You meet a wolf's eyes, and one of you backs down. But with these creatures, according to dozens of reports, the eye contact **lingers**. And it doesn't come with movement. It comes with stillness.

Still body. Still face. Eyes fixed.

One man in Quebec described it like "watching a person who didn't need to blink." He saw it from his treestand, forty yards away, just before last light. He said the eyes weren't wide or glowing. Just deep. Expressionless. Watching. He said it felt like looking into a tunnel. He climbed down and left immediately. Didn't even retrieve his gear.

Another Ontario report from 2017 describes two men hunting on crown land. One stepped ahead of the other at a game trail fork. When he turned back to speak, the second man was standing frozen, eyes locked across the path. "It's watching me," was all he said. Not *it's coming*. Not *what is that*. Just *watching*. They both backed out, guns never raised.

It wasn't fear that made them leave. It was the realization that whatever was out there was **not reacting like an animal**. It wasn't startled. It wasn't cornered. It was calm. Composed. In control.

You can't train for that. You can't explain it to someone who hasn't experienced it. You don't forget it.

It's also why so many of these reports come **without aggression**. The creature doesn't attack. It doesn't charge. It just watches—and then leaves. That's the psychological kicker. Because in the moment, the witness realizes they're **not being hunted**. They're being *evaluated*.

And that's worse.

Some people describe crying afterward. Others get physically ill. A few develop night terrors, insomnia, or recurring dreams about the eyes. I've heard it too many times to ignore: "The eyes were the worst part." Not the size, not the noise, not the impossible fact of what they saw—but how it **looked at them**.

As if it wanted them to know it saw them.

As if it was letting them go.

Part 3: Facial Descriptions—Patterns in Reports

Ask a dozen people to describe a bear and you'll get almost identical answers. Ask them to describe a person, and the details will vary—but the framework stays human. Now ask people to describe a Bigfoot's face.

The patterns in their answers are disturbing in how *consistent* they are—especially from witnesses who've never spoken to each other.

The face, they say, is wide. Not grotesque. Not deformed. Just… wide. The eyes are deep set. The brow is heavy. The nose isn't flat like an ape's—it's broad and human-shaped, though sometimes with flared nostrils. The mouth is wide, rarely open, with lips that are either thick or nearly invisible depending on lighting. And the skin—here's where people pause. Because the skin is usually visible.

These aren't creatures entirely hidden behind fur. Most witnesses say the face is either hairless or lightly covered in fine hair, with patches of exposed skin around the cheeks, brow ridge, and chin. Some say the skin was gray. Others dark brown, even black. One woman said it looked like weathered leather. Another described it as "the color of wet tree bark."

You start to notice that almost none of the serious reports describe a "monster." They describe something *real*. Tangible. Dirty. Alive.

In 2021, a forestry worker north of Manitouwadge came face-to-face with one near a creek crossing. He was in his truck. The creature stood in a clearing less than thirty feet away, staring at him while he fumbled for his radio. He described the face as "half-man, half-something else." Not because it was split, but because it kept shifting in his mind. "I knew it wasn't a man," he said, "but it had *man* in it." He fixated on the jaw—said it looked too strong, like it was made to crack bones. But the eyes were calm. Steady.

Another report from Oregon described the skin around the eyes as "loose," like a human with a tired expression. The brow ridge was pronounced, casting the eyes in shadow, even under bright light. That same report noted small patches of lighter hair near the mouth and temples—like graying.

And these details matter. Because they don't come from movies. They don't sound like someone describing a werewolf or fantasy creature. They sound like someone trying to explain something their brain isn't built to categorize.

The consistency is what drives it home.

These people don't know each other. They've never read the same books. Some can barely get through the interview without shaking. But they all describe a **face that sits between categories** —part human, part other.

Too human to dismiss. Too different to accept.

This is why, in many cases, the face haunts them more than the encounter itself. Because they didn't see a blur. They didn't see a shadow. They saw a *face*. With shape. With skin. With a mouth that could scream and eyes that could think.

And once you've seen that, the world doesn't go back to normal.

Chapter 9: Tracks—What They Tell Us and What They Can't
Part 1: Anatomy of a Print – Size, Stride, Pressure, and Shape

Track evidence is the backbone of most terrestrial animal studies. It's been that way since the beginning—long before game cameras or satellite telemetry, long before night vision or drones. A well-preserved track can tell a field biologist a surprising amount: species, size, gait, estimated weight, even injury or behavioral patterns. And in the world of Bigfoot research, it's no different. In fact, it's where the whole modern search began.

Bluff Creek, 1958. Jerry Crew. A cast of a 16-inch track that set the modern fire. Decades later, the Patterson-Gimlin film, shot in the same area, would show a figure making strides that matched the prints left behind. But this wasn't an isolated event—it was the beginning of a wave. And since then, thousands of tracks have been reported, measured, photographed, and in some cases, cast in plaster.

Some are hoaxes. That's unavoidable. But others… others are too strange to dismiss. Too anatomically correct. Too **functional** to be faked by someone stomping around in wooden cutouts.

The average human stride is around 30–36 inches. Bigfoot tracks often show stride lengths of 48–60 inches—or more. That alone doesn't mean much. A tall man can stretch his stride. But combine that with the **depth** of the prints and the terrain they cross, and a different picture forms. These aren't running strides. These are long, smooth, powerful steps taken **while walking**—something that requires mass, hip flexibility, and leg length outside human norms.

Then there's the foot shape itself.

A real Bigfoot track, when you see one up close, doesn't look like a scaled-up human foot. The arch is absent or extremely flat. The heel pad is broad. The toes are splayed—not squashed together like ours from years of shoe use—and sometimes they shift position slightly across a sequence, as if adjusting for terrain in real time.

This is one of the key anatomical differences that led Dr. Jeffrey Meldrum, a professor of anatomy and anthropology at Idaho State University, to study these prints in depth. Meldrum maintains one of the largest collections of Bigfoot casts in the world—over 300 in total. He's written extensively on **midtarsal flexibility**, a trait seen in apes but not in humans, and how many of the most convincing Bigfoot tracks show signs of a midtarsal break—a kind of foot pivot we don't naturally have. That's not something a prankster would think to fake.

And then there's the matter of pressure ridges.

Some tracks—particularly those left in fine mud or silt—show distinct compression lines along the edges. These are created by the weight and movement of a living foot, flexing as it pushes off. It's extremely hard to replicate these convincingly. A static mold pressed into soft dirt will leave a shape. But it won't leave a **living motion**.

That's how a lot of the best evidence surfaces—not through headlines, but through quiet, incidental discovery. And often, the most telling part isn't the size or shape of the print.

It's the **way it moves** through space.

Part 2: Famous Trackways—Bluff Creek, London, and Beyond

Some tracks stand on their own. Others become legends.

The Bluff Creek trackway is where it all began—or at least, where the modern public fascination took root. In 1958, logging equipment operator Jerry Crew found and cast massive prints in northern California, and a local newspaper coined the name "Bigfoot." What many forget is that this wasn't an isolated track. It was a **sequence**—left in soft dirt over hundreds of feet, unbroken and clean, with consistent spacing and depth. It wasn't one prank print. It was a walking pattern.

Almost a decade later, just upstream from the same area, Roger Patterson and Bob Gimlin filmed a subject now known as "Patty." The prints found at that site—measured, photographed, and partially cast—match the stride and depth seen in the film. In fact, one of those prints later became part of the Smithsonian's reference collection.

But the Bluff Creek prints are just the beginning.

Over the years, several other trackways have emerged that, while lesser known, carry just as much forensic weight.

The London Tracks (1982 – Ohio)

Near the small town of London, Ohio, in 1982, a man discovered a sequence of large prints running along a frozen soybean field just after a snowstorm. What made these unique was the **ice** —the surface was lightly frozen, just enough to show clean compression without breaking the surface tension. That meant the impressions held fine detail without distortion. Local police documented the site, and several tracks were cast. Independent analysts who reviewed the casts years later noted unusual toe movement, subtle heel rotation, and one slightly malformed toe ridge—possibly an old injury.

The Freeman Trackway (1982 – Washington)

In the Blue Mountains of southeastern Washington, researcher Paul Freeman discovered an extensive trackway including **knuckle impressions,** as if the creature briefly dropped to a crouch or used its hands while navigating a slope. Several of the prints showed clear dermal ridges—fine skin texture—under magnification. While Freeman was a polarizing figure, these casts remain in circulation, and even skeptical analysts have acknowledged the depth and anatomy as "biomechanically plausible."

Manitoba Ice Print (2010 – Interlake Region)

One of the most haunting cases comes from northern Manitoba, where a hunter on a frozen trail captured photos of a **single print** in clean ice crust. It was huge—around 16 inches long—but what made it stand out was the **sudden disappearance**. No lead-up. No follow-through. Just one perfect barefoot impression where the trail curved. The ice crust was thin, meaning anything heavy would've broken it. Yet this print compressed the surface without shattering it. The depth profile suggested extreme weight distributed over a wide, flat foot—bare, but not human.

Local Cree elders were consulted. They didn't call it Bigfoot. They called it *Wetikohtin*—a name with no direct translation, but one used for "those who walk differently and do not return the stare."

Consistency Across the Continent

When you lay the casts and measurements side by side—from California, Ohio, Washington, Ontario, British Columbia—you begin to see patterns:

- Lengths typically fall between **14–17 inches**

- Widths average **6–8 inches**

- Toes splay with terrain but remain consistently **five in number**

- Midfoot pressure is broad, suggesting **flat or flexing footpads**

- Depth increases on inclines, especially downhill, which indicates **controlled motion — not bounding**

And most importantly: these prints rarely appear as singles. They show **motion**. Gait. Movement through landscape — not planted for attention, but left unintentionally during travel.

That's what separates credible trackways from hoaxes.

The footprints don't try to be impressive. They try to **go somewhere**.

Part 3: What Tracks Can't Prove — Limits of Footprint Evidence

Tracks are among the most tangible pieces of evidence we have in the search for Bigfoot. They're visible. Measurable. Sometimes even castable. They leave behind real-world data: stride length, pressure distribution, gait analysis. But for all their value, tracks alone can't give us everything. In fact, there are critical limitations that, if ignored, can undermine serious research.

The first — and most obvious — problem is **hoaxing**.

Faking a track isn't difficult. All you need is time, a set of carved feet, and access to soft ground. Some hoaxes are cartoonishly bad — clumsy, repetitive, or anatomically nonsensical. But others are sophisticated. There have been fakes with articulated toes, worn textures, even subtle asymmetry designed to mimic injury or weight shift.

This forces researchers to stay cautious. Even experienced trackers have been fooled. That's why the context around a track is often more important than the print itself: Where was it found? How many prints were there? What was the terrain like? Was the trackway consistent? Did the stride match the depth? What's the local history?

But even when a print is **genuine**, it still doesn't tell the full story.

A footprint doesn't prove species. It doesn't give you DNA. It can't confirm origin or purpose or behavior. It tells you that something *passed through*—nothing more.

That becomes even more complicated in regions with overlapping wildlife. In deep snow or melting ice, **bear tracks** can warp into wide, vaguely humanoid shapes. When a bear's hind foot lands over its forefoot, the result can appear strikingly similar to a large primate print, especially to the untrained eye. The stride will still be wrong. The depth pattern will be off. But from a distance, in the right light, it's a common source of false alarm.

Another issue: **environmental distortion.** Snow melts. Mud dries. Rain softens edges. Ground shifts. A clean print at 8 AM might be warped by 8 PM. That's why fast documentation matters. And why plaster casts taken under poor conditions can do more harm than good—they preserve a *bad version* of the real thing.

Then there's the matter of **absence**. In winter, many researchers report encounters or vocalizations with **no tracks at all**. That's not impossible—frozen crust, hardpan ground, thick leaf litter, or granite outcrops don't always take impressions. But skeptics will rightly ask: if these creatures are real, where are the constant trackways? The migration prints? The seasonal movement trails?

And the honest answer is: **we don't have them**.

That doesn't mean they don't exist. But it does show the limits of print-based evidence. No matter how convincing a cast might be, it's not proof of the creature—only proof that **something** moved through that spot at that time.

For all the science we apply—pressure analysis, 3D modeling, comparative primate foot anatomy—footprints are ultimately **echoes**. They're signs of presence, not identity.

To take the research further, we need more than shapes in the mud.

We need **biology**.

Chapter 10: The Body Problem—Why We Don't Have One (Yet)

Part 1: The Single Question That Always Comes First

In every serious conversation about Bigfoot—whether with scientists, reporters, or skeptics—there's one question that cuts through everything else.

"Where's the body?"

It's fair. It's also loaded. Because beneath it is an entire stack of assumptions: that if Bigfoot is real, we should've found one by now. That death is inevitable. That remains persist. That nature gives up its secrets eventually. But this is where real-world wilderness experience parts ways with armchair logic.

The idea that a body *should* have been found relies on a few conditions that don't hold up in the North American wild.

First: nature does not preserve.

When a large animal dies in the forest, the decomposition process begins immediately. Scavengers move in within hours—ravens, foxes, coyotes, bears. Insects faster. Bacteria fastest. The smell alone draws in everything for miles. Within a week, you're left with scattered bones. Within months, the bones themselves are buried under leaf litter, washed downstream, or chewed apart for calcium.

Even moose and black bears—massive, common animals—are almost never found dead in the wild unless killed directly by humans. I've spent decades in the bush, and I've never stumbled across the fresh corpse of a wild bear. Antlers, maybe. A femur now and then. But full, rotting carcasses? They vanish faster than most people realize.

Second: behavior matters.

We don't know how these creatures treat their dead. And if the face-to-face reports are accurate—if they really are as intelligent as described—then it's not irrational to think they may avoid leaving bodies behind in obvious places. There are primates that bury or guard their dead. Even elephants return to their own fallen. Is it so impossible to imagine that a similarly intelligent, reclusive species might do something we're not accounting for?

Third: remote country swallows everything.

People underestimate how vast and untouched parts of North America still are—especially in northern Ontario, northern British Columbia, the Yukon, Alaska. There are places where no human has stepped in decades. Areas that, even with satellite mapping, remain unvisited, unlogged, and unmapped on foot.

I've hiked regions where GPS failed, where even First Nations maps ended. And I've seen moose skeletons reduced to moss-covered toothpicks in less than a year.

If something ten times rarer than a bear died out there, and the terrain was rugged enough, **no one** would find it.

And last: some evidence may **have** been found—and ignored.

There are occasional whispers in the research community. Unconfirmed stories. Partial bones that "didn't match," and then vanished into lab shelves. Hunters who shot at something they thought was a bear, then left the scene when it stood up and ran. Forestry workers who found skulls and said nothing. It's impossible to verify most of it. But the possibility exists—not that the body was never found—but that it **wasn't believed**.

Because until you believe the creature is real, **you won't recognize its corpse**.

This is the heart of the paradox.

People demand a body to believe. But the belief itself may be the only thing that would allow us to **recognize** one if we ever saw it.

Part 2: Forensic Dead Ends—False Starts and Misidentified Remains

It's not that people haven't tried.

In the past 50 years, dozens of bone fragments, hair tufts, and partial remains have been handed over to universities, state labs, or private researchers—each one submitted with a trembling hope: *maybe this is it*. But almost every time, the results come back the same. **Bear. Deer. Cow. Human. Unknown but non-primates.** DNA degrades fast in the wild. By the time something reaches a lab, it's often too compromised to sequence fully—or too ambiguous to classify cleanly.

Still, there have been moments. Moments where the story could've turned—if only the evidence had held.

1. The "Skookum Cast" (2000, Washington State)

Not a body. But close. During an expedition by the BFRO in the Gifford Pinchot National Forest, researchers found what they believe was a full-body impression left in mud beside a bait station. The shape included what appeared to be buttocks, thighs, calves, and even a heel. The team cast the entire indentation. Some believe it shows a creature lying on its side, reaching for fruit. Skeptics argue it's just an elk bed. The cast still exists. But like many pieces of evidence, it remains in limbo: **too weird to ignore, too vague to confirm.**

2. The Georgia "Body Hoax" (2008)

Two men claimed they had a frozen Bigfoot body in a cooler. Press conferences followed. Photos circulated. For a few surreal days, media across the U.S. ran with the story. Then the truth came out—it was a costume, rubber feet and all, packed in ice. Hoaxers admitted it under pressure. The fallout set serious research back years. To this day, it's used as a reason to dismiss all claims wholesale, despite having no connection to actual field researchers.

3. A Tooth in Northern California (1976)

A private landowner found what he thought was a fossilized molar in a dry wash. The tooth didn't match human or bear. It was large. Heavy. Human-shaped, but far too big. He mailed it to a local university. Months later, he was told it had been "lost during transfer." No records. No analysis. No explanation. Some believe it was a mistake. Others think it was buried on purpose. No proof either way. But the story lives on in regional Bigfoot lore.

4. Hair Samples from the Yukon (1997, 2003, 2019)

Three separate samples of coarse, reddish-brown hair were found clinging to broken branches in different regions of the Yukon wilderness. They were sent to different labs. Results: "unknown primate" in one case, "inconclusive" in the second, and "low DNA yield" in the third. The first result caused a small stir—but the lab refused to officially label it as hominid. Without a full genetic profile, it was shelved. No follow-up was conducted.

5. Unreported Shooting Claims

These stories come from the backchannels—hunters who claim they shot at something they thought was a bear, only to see it stand upright. Others say the creature dropped, but when they returned with help, the body was gone. These reports are usually anonymous. Often emotional. Rarely verifiable. But again—**the pattern** is consistent: a brief glimpse of finality, followed by disappearance. And always, the same regret: *I didn't know what it was until it was too late*.

In each of these examples, the same thing happens. A hint of truth flickers through the static—but never long enough to anchor. Either the sample is lost, the data is too vague, or the source discredits itself before the public can believe.

That's the frustrating part about the body problem.

It's not that we've **never come close**.

It's that we keep arriving one step too late.

Part 3: What a Body Would Actually Mean—Scientific Fallout and Cultural Shock

Let's say someone finds one. Not a blurry photo. Not a print in snow. A real, physical body. Dead or alive. Confirmed. Documented. Sampled. Sequenced. Let's say it happens tomorrow.

Most people assume that would be the end of the debate. But it wouldn't be. It would be the beginning of a **scientific, cultural, and political earthquake**—one that would ripple through multiple institutions, and in some cases, fracture them entirely.

Scientific Institutions Would Splinter

Primatology, anthropology, biology—they'd all be forced to reevaluate major assumptions. Where does this creature fit on the hominid tree? Is it closer to *Homo erectus*? Is it its own genus? Is it relic *Gigantopithecus*? How did it survive? How did it remain hidden? Who missed it—and why?

University departments would scramble. Old papers would be retracted. Tenured academics who mocked the idea would be called out publicly, and researchers who spent decades in ridicule would be invited to speak at conferences they were once banned from.

And all of that would happen before we even begin to **study the body**.

If the DNA is unique, it would unlock decades of new research. If it shares genetic overlap with humans, the ethical storm would begin.

Legal and Conservation Implications

If the creature is real—and if it's *not* human—then it's an **endangered species**, instantly. That triggers global protections under wildlife law. Entire tracts of land—especially in remote northern forests—could become restricted zones. Logging halted. Hunting suspended. Ecological surveys rewritten. First Nations and Indigenous governance might have new leverage in longstanding land disputes. Rural economies could be frozen overnight.

On the other hand, if DNA testing places it **within the human family**, even partially, the fallout gets stranger.

Would it have rights?

Could it be hunted? Displayed? Filmed?

What if the body is juvenile?

What if it's female?

What if it's… alive?

Culture Would React Faster Than Science

The public response would be instant, chaotic, and deeply polarized.

Some would celebrate the discovery. Others would panic. Extremist groups might consider it blasphemy. Conspiracy communities would claim the truth was known all along and withheld. Others would accuse the media of staging it. You'd have protests, pilgrimages, merchandise, government lockdowns, and a new wave of "Bigfoot tourism" hitting every forested region in North America.

You'd also have *every* person with an old story finally coming forward—some for the right reasons, some for the attention. The noise would become deafening.

The Witnesses Would Finally Be Believed—But at a Cost

For the thousands of people who've carried these experiences quietly, for years, the confirmation might feel like relief. Vindication. But it would also reopen emotional wounds. Encounters once buried as trauma would resurface. And those who had been silenced—by employers, communities, even families—would have to decide whether to speak or stay quiet.

And for those of us in the field—those of us who believed, but never had proof—we'd have to start again. Not hunting the myth.

But studying the species.

Because at that point, Bigfoot wouldn't be folklore anymore.

It would be **biology**.

Chapter 11: Hair, Blood, and Tissue—What the Labs Actually Found

Part 1: The Samples Nobody Talks About

While footprints and eyewitness reports dominate most public discussions of Bigfoot, there's another layer of evidence that rarely gets mainstream coverage—because it's harder to dismiss, and harder to understand. I'm talking about **biological samples**: hair, blood, tissue, and other trace materials collected in the field and analyzed in laboratories.

These samples have been submitted, tested, and in some cases, published. But the results—far from proving or disproving anything—have only deepened the mystery. They fall into a strange category: **inconclusive but provocative**.

Take hair, for example. Researchers, hunters, hikers, even police officers have turned in clumps of coarse, reddish-brown or black hair strands that don't match any local animal. No undercoat like bear. No shaft structure like deer. And yet when analyzed, the DNA results come back vague. Most are labeled "human contaminated" or "insufficient sample." Some show mitochondrial DNA that matches *Homo sapiens*, but with no nuclear DNA—meaning the deeper genetic identity can't be confirmed.

This has happened **dozens of times**.

In 1996, a tuft of hair found on a fence near Mount St. Helens was sent to a university in the Pacific Northwest. The analyst reported that the sample had "some characteristics consistent with primate hair" but lacked sufficient markers for a confident species match. That same year, a hair found on a logging road in British Columbia was tested independently by two different labs— both found partial human DNA, but also anomalies in the medulla pattern, the inner core of the hair shaft. Neither could confirm or deny its origin.

And then there's blood.

In 2002, a hunter in northern Alberta reported seeing a large, upright creature near his cabin and fired a warning shot. He claimed the animal fled, leaving behind a blood smear on tree bark. He collected a sample and froze it in a mason jar. Months later, a local biology professor agreed to run a basic analysis. The result: hemoglobin present, blood confirmed. But the sample degraded during shipping. No species match. The man refused to go public. I saw the letter from the lab. It was brief but real. "Unknown origin. Suggest further testing."

And that's where it usually ends.

Because labs are hesitant. They're not in the business of identifying legendary creatures. They want funding, peer-reviewed journals, and clean samples with known parameters. You submit a vial of blood labeled "possible Bigfoot," and they either turn it away or run it under protest. Even if the result is anomalous, they'll write it off as **contaminated**, **degraded**, or **mislabeled**.

Tissue samples are even rarer. Most supposed "flesh" finds turn out to be roadkill, chewed bark, or fungus. But there are a few exceptions.

In 2013, the now-infamous "Sykes Study" at Oxford University aimed to put the Bigfoot legend to rest by running DNA tests on dozens of samples collected from around the world. The results made headlines: most were matched to known animals—horses, cows, bears. But two samples, both from the Himalayas, came back as *anomalous*. Not ape. Not human. Closer to ancient polar bear DNA. The media spun it, but in the footnotes of the report, even the researchers admitted they couldn't explain it.

Back in North America, smaller labs—those working with cryptid researchers or private donors —have kept low profiles. I've spoken with two who said they'd tested hair that defied categorization. One researcher, speaking on condition of anonymity, told me: "It's not bear. It's not deer. It's not human. But I don't know what it is. And if I publish that, I lose my funding."

That's the wall we hit.

Not a lack of evidence.

A lack of **willingness to believe the results**.

Part 2: Mitochondrial DNA, Contamination, and the "Human Primate" Hypothesis

If you spend enough time studying the lab reports surrounding Bigfoot-related biological evidence, a strange pattern emerges. Hair samples, skin cells, and even degraded blood smears often yield the same frustrating result: **mitochondrial DNA matches human**. But no readable nuclear DNA.

To understand why this matters, and why it doesn't solve the mystery, we need to break down what those results really mean.

Mitochondrial DNA—mtDNA—is passed down through the maternal line. It's incredibly stable, easier to extract from degraded samples, and typically the first thing labs sequence. But it doesn't tell the full story of an organism. It's like checking the battery of a phone to figure out what model it is. You might learn a little. But not enough.

Nuclear DNA is the core genetic code—what makes you a human, a bear, or a coyote. That's the holy grail in species identification. But it's also much harder to recover. And here's where Bigfoot samples consistently hit a wall.

Dozens of lab reports, many unpublished, show this same contradiction: **the mtDNA says "human female"**, and the nuclear DNA is either unreadable or shows strange anomalies that cause it to be labeled "contaminated" or "failed."

And that opens a crack.

Because there's a working theory among some researchers—one that most mainstream scientists avoid entirely because of how disruptive it could be. The theory goes like this:

What if these creatures **are part human**?

Not "humans in suits." Not "wild men." But something ancient. Something genetically close enough to share maternal ancestry, yet divergent enough to present completely different physiology—body size, hair, behavior, intelligence, vocalizations, anatomy.

Think about it.

There are already primates with over 98% genetic similarity to humans. If a relic hominin survived in isolated environments—adapted, elusive, nocturnal, and socially secretive—it might look a lot like what's being described in these reports. If such a species existed, its mitochondrial DNA could easily trace back to *Homo sapiens* or a common ancestor.

In 2012, a controversial study led by Dr. Melba Ketchum tried to put this theory to the test. Her team analyzed dozens of hair samples, many of them submitted by independent field researchers from North America. The conclusion: **mtDNA showed human**, but nuclear DNA showed an "unknown hominid." The scientific community rejected the paper—not because the results were impossible, but because the process lacked peer review and proper replication.

Still, the data remains.

And the implications are seismic.

If even one of those samples is authentic, it suggests that Bigfoot isn't an undiscovered ape. It's something **closer**. Something that shares part of our code.

And that may explain why so many labs hesitate.

Because if you sequence something that's part human, you enter a legal and ethical minefield. Tissue rights. Species classification. Conservation laws. It's no longer just a mystery—it's a **discovery that redefines us.**

And most labs? They aren't ready for that.

That's why so many samples are quietly shelved. Why "human contamination" has become a kind of catch-all for anything that doesn't make sense. It's easier to say "error" than to say "unknown primate."

But behind the scenes, the whisper is getting louder. Researchers are comparing notes. Geneticists are starting to look twice.

Because the numbers don't lie.

And they point to something walking the line between legend... and legacy.

Part 3: What We'd Need for Proof—Protocols, Chains of Custody, and the Gold Standard Sample

For all the anecdotes, all the strange test results and "unknown" labels, the truth is this: until we get the right kind of sample under the right conditions, **none of it will convince mainstream science**. The mystery isn't enough. What's required is a clean, undeniable sample—and a flawless trail of how it was obtained.

In other words, we don't just need a hair or a drop of blood.

We need a **forensically defensible biological specimen**, collected under strict protocols, with complete documentation and zero possibility of tampering.

Here's what that would actually look like:

1. Collection Under Observation

The first problem with most evidence today is that it's found after the fact. No camera. No witness to the sample's origin. That leaves gaps—gaps that skeptics are happy to fill with doubt. A real breakthrough would require a sample **taken immediately after a verified encounter**, with photos, video, and timestamps showing its exact location and context. Ideally, you'd have it happen on camera—thermal, night vision, high-res. A drone overhead wouldn't hurt either.

2. Chain of Custody

This legal term matters more than most people realize. From the moment a sample is taken— hair, blood, tissue—it must be **sealed, logged, and handed off with documentation**. Every person who touches it needs to sign off, note its condition, and be accountable for its transfer. No shortcuts. No back-of-the-truck stories. Just clean, traceable evidence.

That's how murder trials work. That's how endangered species are identified. And that's how Bigfoot will need to be proven.

3. Sterile Handling and Environmental Controls

The biggest failure in past sample tests has been contamination—sometimes from the person collecting it, sometimes from the lab. To avoid this, the sample has to be taken with **gloves, sterile tweezers, and stored in containers designed for biological preservation**. Refrigeration matters. So does humidity. So does the speed of delivery to the lab. The moment you throw something in a Ziploc bag and stick it under your truck seat, the chain breaks. The sample is compromised.

4. Independent Testing at Multiple Labs

Even if you do everything right, no one will accept a single lab's results. That's the world we live in now. To be believed, a sample would need to be divided and tested **independently by**

multiple institutions, preferably in different countries. Each lab would need to publish its methods, results, and full DNA sequencing. And ideally, those results would match.

Only then would the scientific community pause.

Only then would the "unidentified" label become something more.

5. Correlated Evidence

And finally, the sample needs **context**. A print nearby. A video. A recorded vocalization. A detailed report with environmental conditions, GPS location, time of day. This isn't about overkill. It's about creating a **case**. Because even if the DNA is clean, the first accusation will always be: *"Where did it come from?"*

This is the standard. And it's hard to meet.

But it's not impossible. In fact, it's already close. More researchers are moving toward forensic protocols. More are working in teams, documenting better, preserving smarter. The era of blurry photos and plastic bag hair samples is ending.

If someone gets the right sample under the right conditions… it will change everything.

Not because it's the first evidence.

But because it's the **first that no one can ignore**.

Chapter 12: Why They Vanish—Predators, Scavengers, and the Disappearance of the Dead

Part 1: Vanishing Acts—Why Remains Disappear in the Wild

People have this image in their heads: that if something dies in the woods, it just lays there, waiting to be found. But if you've spent any real time in the bush—deep bush—you know that's not how nature works. In the North, the wild doesn't just forget bodies. It **consumes** them.

I've walked game trails in Northern Ontario where you can smell something dead on the wind—but never find the source. It's not that it's not there. It's that you're **too late**.

When an animal dies in the wild—any animal—it sets off a sequence. First come the insects. Blowflies, ants, wasps. They arrive within hours, sometimes minutes. They lay eggs in the eyes, the mouth, the soft tissues. By day two, the carcass is alive with larvae. Then the scavengers follow. Ravens. Coyotes. Bears. Wolverines, if you're far enough north.

They don't wait for rigor mortis. They tear flesh loose while it's still warm. Limbs are dragged. Bones separated. Skin devoured. Within three days, the scene that might've shocked a hiker becomes almost unrecognizable.

Then there's the weather.

Rain flattens. Wind buries. Snow covers and holds everything in stasis until spring—at which point rot returns, faster than before. Leaf litter swallows what's left. Roots pull bones downward. Moss and fungus grow over femurs. In time, the only evidence of a once-living body might be a patch of discolored soil and a curve of fractured bone mistaken for a tree root.

People ask: why haven't we found a Bigfoot skeleton? I ask back—have you found a bear skeleton in the wild? A fresh moose skull? A cougar carcass? Because I haven't. And those animals number in the tens of thousands.

Whatever Bigfoot is, it's rare. The sightings suggest that. The spacing of trackways suggests that. The sheer elusiveness of its behavior suggests that. If they die like other large mammals, they're dying in places **no one goes**, and they're disappearing like everything else does in the boreal wild: fast, thoroughly, and with no marker.

And that's if the body is even **left out in the open.**

Some field reports suggest otherwise.

There are rare but consistent whispers from northern regions—Ontario, British Columbia, Alaska —describing what appear to be **protective behaviors**. Witnesses who've seen figures standing over something prone. Others who've heard vocalizations and movement patterns that suggest *groups* gathering at a single point. One case I reviewed from near the Albany River described a "horrible smell" in the undergrowth, followed by silence, then rapid movement away—too fast, too heavy to be deer. No signs left behind. Just a sickening musk and a churned patch of ground.

Were they tending to a dying member of their group?

Was it territorial defense?

We can't know. But we can observe what happens when **bears** die. They often retreat to brush-covered dens or secluded patches of rock. They die alone, in silence. Finding them is rare. Finding the bones afterward? Almost impossible.

If Bigfoot is more intelligent—if it understands injury and death—then the possibility exists that it doesn't just **let** its dead rot in the open.

It might bury. Cover. Conceal.

And in that case, we're not looking for remains anymore.

We're looking for **intent**.

Part 2: How Nature Erases—From Predator Patterns to Soil Biology

If you want to understand why we haven't found a body, it's not enough to just walk the woods. You have to understand how the woods **work**. How nature doesn't just allow death—it **recycles** it, fast and efficiently.

Start with predators. When something dies, the local food web doesn't hesitate. Every scavenger species—from beetles to black bears—has evolved to exploit decay. They arrive with precision timing. Ravens are often first, circling high above areas where nothing else has stirred. Their memory of food sources and vocal communication systems alert others—coyotes, foxes, even smaller carnivores like martens or lynx.

Once a carcass is breached, a feeding frenzy begins. Skin is pulled back. Muscle is stripped. Internal organs go first, usually within hours. Limbs are removed next. Scavengers don't dine with etiquette. They tear, break, scatter. If the animal was large—moose, bear, or hypothetically a Bigfoot—the body would be disassembled within days. What remains are bones. But even those don't last long.

Rodents chew them for calcium. Rain softens them. Wind moves smaller fragments. Large predators will carry off skulls or femurs for sport or nesting. Insects burrow beneath.

Then the soil takes over.

Forest floor biology is its own form of time. Fungi and bacteria break down tissues faster than the naked eye can track. Earthworms mix it into the dirt. Mycelium networks digest the carbon. In wet, acidic environments—like peat bogs or black spruce forest—bones can degrade in a single season. I've stepped in moose tracks from the year before that were **deeper** than the remains of an actual dead animal left for just two months.

You might find a piece of tooth. A piece of cracked vertebrae. But even then, unless it's **identified instantly**, the forest reclaims it. Mud flows. Leaves fall. Spring melt shifts everything. In five years, it's like it was never there.

And that's assuming the animal died in the open.

Many creatures don't.

We know black bears will crawl into brushy thickets, under ledges, or into burrows to die in isolation. Some believe wolves do the same. If Bigfoot has a territory, if it operates in small family units, then it's not hard to imagine it seeks **cover and concealment** during death. The woods are full of nooks—limestone fissures, storm-blown root bowls, sinkholes filled with moss. Once something falls in, it stays hidden unless someone stumbles upon it by pure chance.

It gets stranger when you factor in seasonality.

In winter, the snowpack preserves. But spring melt floods. Bones shift and slide. You might track something one day and find a melt line the next with no sign of what came before. And in summer, the leaf canopy blocks aerial searches, while ground growth obscures even large objects.

And that's the **visible forest**.

Then there are underground factors—acidic soils, high microbial activity, and animals that dig. Porcupines. Groundhogs. Badgers. Even foxes will drag bones into dens. Anything not exposed to air and sun begins to break down from the inside out.

So if you're out in the bush looking for a skeleton, ask yourself this:
Have you ever found the full remains of a coyote? A wolverine? A lynx?

Probably not.

Now imagine something rarer, smarter, and more elusive. Dying in terrain no one reaches. And being absorbed into an ecosystem that doesn't wait.

Chapter 13: Northern Case Files—Witnesses from Ontario's Deep Interior

Part 1: Cottage Roads, Winter Prints, and Voices in the Trees

What sets Northern Ontario apart isn't just the size of the land. It's the silence. The stillness. And the way things seem to happen when no one else is around to see them—but someone always does.

One of the most credible reports I've come across came in December 2013, from a family visiting their cottage for the winter holidays. The snow had just started—two inches at most. They were driving slowly down the unplowed road when the father spotted a trackway cutting across the surface.

He stopped the car. They all got out.

What they saw was simple but chilling: **a series of enormous footprints**, around 15 inches long, 8 inches wide, moving in a perfectly straight line. The prints followed the road for some distance before veering off into the bush. There were **no drag marks**, no tailing, no signs of double-stepping or melting distortion. Just clean, crisp, human-shaped impressions—*only far too large to be human.*

The family filmed it. Their voices on the recording are calm at first—curious, unsure. But as they walk the trackway, something changes. Their tone gets lower, quieter. You can hear one of them say: *"That's not a bear."* Another adds, *"There's no claws."*

Eventually, they measure one of the prints. Fifteen inches. They count out over a hundred prints in total before the footage cuts off.

They didn't submit the video to YouTube. They didn't call the news. They sent it to a small Ontario-based Bigfoot investigator, someone who keeps records quietly and doesn't push for media exposure. That's where I saw it. Not viral. Not dramatized. Just **real people** trying to make sense of something they knew they weren't supposed to find.

That's the thing with these Ontario cases.

They're often quiet. Straightforward. Sober.

Like the woman near Hearst who saw a shape crouched near her barn at dusk, and then saw it **stand up**—taller than the roofline, broad and completely silent. She didn't scream. She didn't grab a camera. She closed the door and locked it. And then she called her son.

Or the two snowmobilers west of Cochrane who heard "something massive" pacing them just out of sight along a frozen hydro cut. They stopped their engines. It stopped. They revved. It moved again. One of them tried to ride up the side trail toward the sound—nothing. Not a footprint. But the feeling, they said, was *like being followed by something that knew the trail better than they did.*

And then there's the camp report from outside Chapleau.

Late summer. Group of teenagers at a firepit. No alcohol, no phones, just bush and sky. One girl stepped away to use the outhouse, came back pale and shaking. She said she saw "a man crouched behind the woodshed." Only it wasn't a man. It didn't move like one. She couldn't see its face—just the shape, and the feeling that it knew **she** was there before she noticed it.

When they checked, the ground behind the shed was flattened. No tracks. But a faint smell—musky, wet earth, slightly sweet—hung in the air for over an hour.

She never went back to that camp.

These aren't campfire stories.

These are **consistent**, **quiet**, and often **never reported** beyond a friend, a relative, or a researcher like me.

There's a rhythm to them.

A shape.

And the farther north you go, the more you realize—it's not about proving it to the world.

It's about understanding **what kind of world** this really is.

Part 2: Behavior in the Boreal—How Encounters Change the Pattern

In Northern Ontario, the encounters don't just follow a pattern—they *create* one. Over time, the more I listened to witness accounts, the more I noticed a subtle but unmistakable trend: **behavior changes after sightings**. Not the creature's. The witness's.

Something shifts in them.

And it's not always fear.

Sometimes it's caution. Sometimes curiosity. Sometimes it's something quieter—like reverence, or the sense of having seen something sacred they're not sure they were supposed to witness.

But let's begin with the behavior **before** the encounter.

Almost every report I've collected in this region involves someone going about their life **normally**. Cottage visits. Hunting trips. Work on logging roads. Checking traplines. Nothing out of the ordinary. Then something strange happens—**a sound, a movement, or a sense of being watched**.

They pause.

They look.

And what they see—or hear—is something that **doesn't belong**.

It might be a figure moving upright across a ridge. It might be a vocalization—deep, long, resonant—coming from the opposite side of a lake. Or it might be just a feeling, heavy in the chest, like something just stepped through the silence that shouldn't be there.

From that point on, the behavior changes.

Take the trapper who worked the edge of a black spruce line east of Timmins. For years, he ran the same trail system—seasonal, productive, no problems. But in 2020, he found several snares twisted into shapes no animal could've made. Not broken. Not sprung. Just **tied into loops and knots**, hanging from tree branches three feet off the ground.

He didn't hear anything.

Didn't see anything.

But he changed his route.

A week later, new tracks appeared along the side trail. Huge. Flat. No claws. Just a single line of prints that mirrored his—same entry point, same exit, no deviation.

He didn't go back that season.

Or the family that stopped hunting their favorite moose corridor north of Gogama after hearing "three vocalizations at dawn"—a rising whoop that echoed across the river followed by two short, guttural barks. No wolves. No coyotes. The sound didn't fit anything they knew.

The father said: "It was like something calling roll. Like it wanted to know who else was awake."

He never hunted that spot again.

It's not fear in the classic sense.

It's **a recognition**. A shift. The forest isn't just backdrop anymore—it's *watching*. The silence becomes weighted. The trees, more than scenery. You don't leave trash behind. You don't yell when you chop wood. You listen more.

That's the real footprint left behind in many of these cases.

Not in snow.

Not in soil.

But in **people**.

In how they move.

How they look into the bush before stepping off a trail.

How they lower their voices when the sun drops past the tree line.

Because once you've seen something in the forest that **shouldn't exist**, you never go back to thinking the forest is empty.

Part 3: Cold Weather, Quiet Roads — Why Northern Sightings Stick Differently

There's something about Northern Ontario in winter that makes sightings *stick* — not just in memory, but in the landscape itself. The snow, the silence, the long hours of darkness… they don't just create the perfect conditions for encounters. They create the perfect conditions for **documentation**.

No foliage to hide footprints. No summer wind to swallow sound. No crowds to drown out an experience with disbelief.

Just you, the woods, and what's walking through them.

I've seen print lines in snow that stretch for thirty or forty paces. No signs of collapse. No drag marks. Just massive, flat-bottomed impressions with wide toe splay, moving in straight lines impossible for a human to mimic without deliberate effort — and no reason to.

But it's the stillness that drives it home.

Take the report near Smooth Rock Falls in January 2019. A man driving an old pickup along a plowed service road spotted something dark ahead, just at the edge of the headlights. He slowed. At first he thought it was a moose crossing — wrong shape, though. Too upright. The thing took two steps, then cleared the ditch in a single motion. He stopped the truck and waited. His breath fogged the cab windows. The radio was off.

He said the silence after the figure moved off was the loudest he'd ever heard in his life.

Not even the engine ticking.

Just a memory hanging there — impossible to forget, and equally impossible to explain.

Or the snowshoer west of Foleyet, who spotted prints following hers along a treeline. She hadn't heard anything. But when she turned around on her loopback, the tracks were there — fresh, slightly overlapping hers. She said they were "like a man's, but flat, and deep. And there were only two at a time, not a pair like someone walking normally."

She followed them for a few minutes before they cut off into dense tamarack. Her exact words? *"I didn't feel scared. I felt… like I wasn't supposed to be there anymore."*

And that's something that comes up again and again in cold-weather reports.

Not panic. Not primal fear.

Displacement.

A sense of having walked into someone else's space.

People who've lived up north long enough know the feeling of being watched. They know when a bear is too quiet. When wolves are too coordinated. When the trees bend the wrong way in the wind. So when they say something felt "off," I listen.

In the summer, you can second-guess shadows. Blame sounds on wind, insects, tourists, ATV noise. But in the dead of winter, in that frozen hush where every noise carries for miles, the unusual becomes *clearer*, not harder to explain.

And it's not just the environment that makes it memorable.

It's the isolation.

Most of these witnesses weren't looking for anything. They weren't believers. They were alone, or with one person, with no one to convince and no reason to lie. They weren't filming. They weren't measuring. They were just **out there**—until something reminded them that they weren't alone.

That's what makes northern winter encounters different.

They don't need proof.

They're already **imprinted**—in snow, in silence, and in the minds of those who were never expecting to believe in anything.

Chapter 14: Behavioral Threads—Patterns That Defy Coincidence

Part 1: Repeating Moves Across Vast Distances

Ask ten witnesses what they saw, and you'll get ten different stories. But ask them **how it moved**, or **what it did just before it disappeared**, and a strange thing happens. The stories start lining up.

It's not the same event. Not the same year. Often not even the same province or state. But the behaviors repeat like echoes through a hundred different woods.

Let's start with the **parallel pacing**.

This is one of the most consistent—and unsettling—patterns I've encountered. A person is walking a trail, usually alone, when they hear **footsteps in the bush**. Not crashing. Not running. Just walking. Slowly. Step for step. Matching their pace.

They stop. The steps stop.

They walk. The steps resume.

But no matter how hard they look into the trees—no movement. No shape. Sometimes a faint shadow. Sometimes nothing.

I've collected these stories from as far north as Attawapiskat and as far south as Georgia. Same pacing. Same deliberate mirroring. Same *presence without appearance*.

It's not just eerie. It's tactical.

Then there's the **stone-throwing**.

Somewhere along the line, skeptics started calling this a hoax indicator—like it was too obvious, too performative. But try to find a black bear that throws a rock. Try to find a hunter that wants to toss a stone **toward their own cabin** late at night. Try to explain why a fist-sized rock lands ten feet from your firepit, and there's no one in sight, no trail access, no reason for it to happen— except that it's **meant to drive you back**.

I've had witnesses describe these stones landing in patterns: *first one*, a long pause, *then two fast ones*, like punctuation marks. It's not random. It's not chaotic. It's *intentional*.

Some patterns are even more subtle.

Like the way they circle.

There are accounts—dozens—of campers who hear bipedal movement around their tents in a **slow, expanding loop**. First near the water. Then behind the trees. Then closer, behind the tent. Always one step too far away to see, always just out of reach. As if drawing a spiral until something changes: the wind shifts, the fire pops, someone speaks. Then silence.

And it's not just at night.

Daylight encounters have their own behavioral repetitions. Sudden freezing. Crouching near creek beds. Standing behind trees, with only part of the figure visible—shoulder, leg, half a face. As if giving you just enough to know what you saw, but never enough to prove it.

That's not coincidence.

That's **patterned evasion**.

One of the most striking consistencies is what I've come to call the **decision stare**. It shows up in cases where the creature is seen at close range—usually under 50 feet. The witness locks eyes with it, and for a moment, it doesn't move. It just looks. Not like a deer. Not like a bear. Something else.

One woman told me, "It looked like it was deciding whether or not to let me keep standing there."

That line has been echoed almost verbatim by others. Different provinces. Different decades.

The actions are different. But the **intent** feels shared.

It's as if they have a playbook—one shaped by generations of staying hidden. Watch. Mimic. Distract. Move. And above all: **never be fully seen**.

These aren't just stories.

They're fragments of a behavioral map—one we've been collecting for seventy years without realizing it.

And the more we overlay them, the clearer the shape becomes.

Part 2: Sound, Stone, Shadow—Communication or Control?

One of the most misunderstood elements of Bigfoot encounters isn't what people *see*. It's what they *hear*. And more importantly, **what those sounds seem to do**.

We like to think of animals as reactive—responding to threats, defending territory, hunting prey. But the reports I've gathered across northern Ontario, and from researchers in Washington, British Columbia, the Adirondacks, and even the Ozarks, point to something else entirely: **sound as strategy**.

Let's start with **wood knocks**.

On the surface, they seem simple—two hard cracks, like someone smacking a bat against a tree. But listen to the context. A lone hiker enters a new area. Nothing visible, no known danger. Then: *crack… crack… silence.*

Sometimes a single knock. Sometimes three, spaced unnaturally evenly. These sounds often come from **opposite directions**, as if flanking or bracketing the human presence. In one case near Lake Wanapitei, two knocks sounded from both sides of a ravine trail within five seconds of each other—clear, resonant, deliberate.

No echoes. No repeating tree creaks.

Just… presence.

Is it a warning? A signal? A kind of territorial sonar?

We don't know. But they aren't random. And they aren't the only sounds that don't fit natural wildlife behavior.

There are also **vocalizations**.

Howls are reported most often at night—deep, long, and pitched in a way that splits the air. I've had hunters tell me it stopped them mid-step. One logger near Red Lake said it "felt like it came from the ground and the sky at the same time." Coyotes don't do that. Wolves don't sustain like that. Bears can bellow—but what these witnesses describe is closer to *language*, or at least **patterned phrasing**.

There's also a report pattern that crops up when people get too close.

Rocks thrown into water. A stick snapped loudly, once, like a rifle crack. Not just to startle—but to **redirect**.

I interviewed a couple who were hiking near Latchford. They came across a creek with unusually trampled edges—flattened cattails, deep depressions. As they approached, a rock the size of a baseball *plunked* into the water between them and the spot. No movement visible. Just the splash. The man froze. The woman said it was like someone was saying: *"That's far enough."*

It worked.

They turned around.

Which brings us to **shadow plays**.

I don't mean hallucinations or flickers of light. I mean the way figures have been seen **half-visible**—just enough to be noticed, never enough to be confirmed. Standing behind trees. Stepping out for one second, then gone. In full daylight. Always too far for a phone camera. Always perfectly timed to vanish just before you commit to approach.

It happens too often to be clumsy chance.

We're not just dealing with an animal that avoids humans.

We're dealing with one that knows how to **manage attention**.

To appear, but not too much.

To threaten, without violence.

To create a situation where the human retreats **on their own**, without ever having proof to show for it.

That's not just intelligence.

That's **control**.

And it's consistent. Across time. Across countries. Across environments.

If this were a deer or bear, those encounters would end with sightings—or collisions. Not stone-throws, flanking sounds, and evasive silhouettes. Those aren't responses.

They're **tactics**.

The more we study these behaviors, the more one thing becomes clear:

We may not understand their language.

But they clearly **understand ours**.

Part 3: Near the Firelight—Why Some Sightings Happen When We Let Our Guard Down

Campfires create a kind of false confidence. They push the dark back a few feet, give us warmth, give us light. But they also **make us visible**. They dull our hearing. They hold our attention. And most importantly—they tell everything out there exactly where we are.

That's when many of the closest encounters happen.

You'd think sightings would peak during hikes, long-range tracking, trail cameras. But some of the most unnerving reports come from people doing nothing more than sitting around a fire—talking, eating, relaxing, unaware they're being watched.

There's a particular kind of stillness that settles over the woods when something is watching. The bugs quiet down. Wind seems to pause. And then there's the sound—a twig, a shift in leaves, a footfall—but it always comes **just past the firelight**.

Too far to see. Too close to ignore.

One case that stuck with me came from a family west of Thunder Bay. It was summer, dry and cool at night. They were sitting by the fire, kids roasting marshmallows, the parents drinking tea. Around 10 p.m., they heard a long exhale from the treeline.

Not a bear snort. Not a deer wheeze.

It was **deep**, measured, and low—like a slow breath forced through a massive chest. The father shined a flashlight toward the sound. Nothing. Then a rock landed ten feet behind them, near their picnic table.

No one said anything for a full minute.

Then the mother whispered: *"Don't scream. Let's go to the cabin."*

They didn't run. They just walked. And behind them, in the woods, they could hear something **matching their pace**.

Another report, from near Mattawa, described a campfire group hearing steady knocking sounds. One knock every twenty seconds—always from a different direction. It circled. First north. Then east. Then south. Each knock closer. When one of them stood up and shouted, *"We know you're out there!"*—everything went dead silent. Like even the wind had frozen.

That's the difference firelight makes.

It creates a **false center**—a sense that nothing can touch us as long as we're inside the circle. But in reality, the firelight is like a spotlight on a stage. And the forest? It's the audience. Watching. Waiting. Deciding.

Several experienced outdoorsmen have told me the same thing: if something wants to approach, it will do so when you're relaxed. When your guard is down. When you're **talking louder than you should**, focused on your coffee, or distracted by a camp chair squeak.

That's when the breath at your back catches your attention.

That's when the hairs rise on your arms, and you suddenly realize the woods have **gone completely quiet**.

And by the time you turn around—there's nothing there.

Not because it wasn't.

But because it knew exactly **how long to wait**.

Chapter 15: The Evasion Blueprint—Why They Avoid Cameras, Roads, and Open Areas

Part 1: Anti-Camera Behavior and the Failure of Modern Tech

I've placed dozens of trail cams in Northern Ontario. I've camouflaged them, masked the scent, run them on thermal-only triggers, even buried them in brush and pointed them toward bait zones. I've caught moose, bears, wolves, hikers, illegal loggers, and once—a white moose passing through snow at dusk. But never anything resembling a Bigfoot.

And I've been in areas with sightings within 48 hours of camera deployment.

Some researchers like to dismiss this as bad luck, or low population density. But too many of us —those who spend real time in the field—have seen the same thing: **everything shows up on camera** except the one thing we're actually looking for.

It's become something more than coincidence. It's a trend. A pattern. A silence so exact, it begins to feel deliberate.

Let's look at the mechanics.

Trail cameras emit **infrared light**. Even when set to no-flash, they give off a faint high-frequency whine—above our range, but well within the hearing of sensitive animals. I've had wolves glance directly into a cam from twenty feet and flinch away. Bears will sometimes destroy them outright. But in almost every case where a cam has been placed **following a Bigfoot sighting**, the result is the same: no image. No audio. Just the usual night forest activity —and a sense that something knows the cam is there and avoids it.

So what's going on?

Some suggest their eyesight is better than ours—particularly in low light. Others believe they hear the battery hum or detect electromagnetic signals. But whatever the mechanism, **they avoid them**. Not just randomly. Strategically.

Even the placement matters. There are reports where individuals have approached a bait zone and **stayed just outside the camera's cone**—as if they somehow understand its field of view. That's not the behavior of a bear or cougar. That's something else. Something aware.

More than once, I've heard from hunters who put up a cam, then had **no wildlife at all** in the area for days. As soon as the cam was removed? Elk. Moose. Coyotes. Activity returned.

Now apply that to something smarter.

Apply that to something that's been evading modern humans for decades.

And apply it to an animal that may have been **paying attention to us** long before we ever noticed it.

Then there's the AI factor.

Recently, several platforms have tried using AI-assisted image sorting to scan through massive collections of trail cam photos. Thousands of hours of footage. Tens of thousands of stills. The result? Nothing new. At least not yet.

But some researchers have quietly reported **false negatives**. Shadowed figures at the edge of frame. Movement caught between two stills. Audio spikes without corresponding motion triggers.

AI is only as smart as the patterns it's told to recognize.

And Bigfoot, by definition, doesn't **fit any known pattern**.

It's not just an evasion of cameras—it's an evasion of classification.

The longer this behavior continues, the more likely it is that we're dealing with an animal that **recognizes observation itself as a threat**. One that doesn't just avoid humans, but avoids **being recorded**. It's as if the act of capture—visual, auditory, technological—is something it understands and actively subverts.

Not supernatural.

Not mystical.

Just a natural-born escape artist living in the only place we still underestimate: the deep forest.

And that's why our technology fails.

Because it's not just us looking for them.

They're watching back.

Part 2: Why Roads Are Dangerous, and How Bigfoot Learns to Avoid Them

Roads are a human invention, but the fear of them is **natural**—especially for something trying to remain unseen.

Most Bigfoot sightings that occur near roads are what I call *mistake moments*. A crossing. A brief silhouette caught in headlights. A figure crouched near a ditch, suddenly rising and vanishing in a single bound. These moments are fast, uncertain, and always without clear follow-up. No extended views. No prolonged observations. Just a flicker, and it's gone.

The pattern suggests something critical: **they don't want to be near roads**.

And when they are—it's almost always accidental, or at night.

Think about it biologically. Roads are:

- **Loud**

- **Bright**

- **Unpredictable**

They bisect terrain in ways that defy the natural flow of animal movement. They're filled with foreign smells—oil, rubber, antifreeze, cigarette smoke—and the mechanical rattle of machines that can't be tracked by scent or sound. For most wild animals, roads are something to cross quickly or avoid altogether.

But with Bigfoot, it goes a step further.

There are reports where individuals were spotted **just before** crossing, only to vanish backward into the woods. There are others where a witness driving late at night saw something *about to step out*, but instead freeze and step back.

That's the key.

They don't freeze like a deer. They **decide**.

It's like they've learned: roads = exposure.

Many researchers believe these creatures pass knowledge down generationally—territorial patterns, vocalizations, avoidance tactics. And if that's true, it makes sense that **avoiding roads** would be among the first lessons. Roads lead to humans. Humans lead to danger.

And then there's the pattern of **daylight avoidance**.

Roadside sightings almost always occur between dusk and dawn. Even logging roads and remote two-tracks—if there's a chance of traffic, they're used during low light. That's not random. That's strategic.

In one case near Wawa, a man hauling a snowmobile on a flatbed watched something move along the shoulder of the Trans-Canada at 4:30 in the morning. It was upright, fast, and dipped down into the ditch without even pausing to check traffic. He slowed and looked. Nothing. Just disturbed snow and a long, curved print on the inner bank of the ditch wall.

Later that year, another witness in the same region saw a nearly identical figure cross a gas pipeline trail at dusk. But again—never during peak daylight. Never when activity is at its highest.

The theory that they use roads at night as **navigation corridors** has some weight. Roads cut through territory. They offer long lines of sight. For a creature that's smart, strong, and primarily nocturnal, the road might be a tool—but only when it's sure it won't be seen.

And sometimes, they get it wrong.

That's when we get the sightings.

That's when the stories start.

But for every one we hear, how many crossings happen at 2 a.m., in fog, in silence, without headlights to catch them?

How many moments happen when the driver blinks or reaches for a coffee?

We assume we'd see them if they were out there.

But we forget how often we miss the **ordinary**.

So imagine trying to see something that doesn't want to be seen—and **knows how not to be**.

Part 3: Open Ground Is a Death Sentence—Movement Through Cover and Why We Rarely See Them Run

You can tell a lot about an animal by the way it moves. Not just its stride or speed, but the *path* it chooses.

And what we've learned from decades of sightings, prints, and close-call encounters is this: **Bigfoot doesn't like open ground. At all.**

This may seem obvious at first. After all, plenty of animals stick to cover—deer bed down near brush, wolves run treelines, bears hug riverbanks beneath canopy. But Bigfoot takes it to a level that goes beyond instinct.

They don't just prefer cover. They treat **exposure like a threat**.

Ask any witness who's seen one moving across a field or powerline cut—what struck them most wasn't just the creature's size or speed. It was how quickly it **got out of view**. A few strides. Then gone.

Almost every open-ground sighting ends the same way: *It moved into the trees so fast, I couldn't believe it was gone.*

Not running. Not fleeing. **Exiting.**

There's a calculated economy to the way they move. When forced into the open, it's usually:

- **In a straight line**

- **Without hesitation**

- **With explosive speed**, and often at an angle that minimizes visibility

Then, back into cover.

It's like they plan their paths before stepping out.

Hunters and trappers in Northern Ontario know how to read terrain for animals on the move. Travel corridors, edge habitats, funnel points. But in Bigfoot reports, you start to hear something else: *"It was like it knew exactly where to go before it even moved."*

One man told me he saw a figure cut across a gravel pit just outside Kirkland Lake. He only saw it for six seconds—long enough to estimate its height, weight, and stride—but what stuck with him was this: *"It didn't run. It just… vanished. I realized later it had dropped into a gully that I hadn't even noticed until I walked the site myself."*

That's terrain knowledge. That's intent.

We've also noticed that print trails—when we actually find them—almost always move **along the edge** of terrain features. Ditches. Brush lines. Ravines. They're not meandering. They're directional, efficient, and always making use of the best possible cover.

Why?

Because open ground is risk. And risk, in the world they seem to inhabit, means **death**.

It's not just human presence they avoid—it's **detection**.

Open terrain means visibility. Visibility means attention. Attention means consequences. And these creatures seem built around the idea of *avoiding consequences at all cost.*

Even when they're seen, they never panic. They don't thrash through brush or crash through branches. They flow. They glide. They move like something that knows the land better than we do.

And that's the point.

We imagine them as primitive.

But what if they're not?

What if the reason we rarely see them run is because they **don't need to**?

Because every step, every turn, every exit has already been mapped—silently, instinctively, over a lifetime spent learning a land we've only begun to scratch.

Chapter 16: Name Games—Why 'Bigfoot' Doesn't Tell the Whole Story

Part 1: Words from the Woods—The Many Names of the Same Mystery

Before the word "Bigfoot" became part of pop culture, before grainy footage and documentary voiceovers, people were already talking about *something* in the woods. Something tall. Silent. Hidden. Watching.

They just called it by different names.

And that's where things get interesting.

Across North America, almost every region has its own name for the creature we now lazily group under the Bigfoot banner. But these names aren't just folklore labels—they reflect subtle differences in **behavior**, **appearance**, and even **habitat range**. In fact, the linguistic trail might be one of the most important—and overlooked—clues we have about what we're dealing with.

Start in the Pacific Northwest, and you'll hear the name **Sasquatch**, adapted from the Halkomelem word *Sésquac*, which loosely translates to "wild man of the woods." But even within that language family, variations exist. Some dialects describe a guardian spirit. Others refer to a flesh-and-blood being. Some describe it as a giant with a face "not like a bear," others as a more elusive shape-shifter.

Move into British Columbia and you'll hear **Steta'l**, **Bukwas**, and **Dzunukwa**—each with different attributes, from child-stealing forest beings to hairy hermits that speak through knocks and howls.

In Northern Ontario, things shift again.

Among the Ojibwe, there's **Sabe** (pronounced Sah-BAY)—a being considered real and physical, not mythical. Often described as a forest protector. Not evil. Not helpful. Just there. Present. Watching.

Witnesses who've used that name usually do so with a kind of weight in their voice. Not superstition. Not fantasy. **Recognition**. The sense that they're describing something *they know exists*.

And then there are the more localized names—terms passed down through families or rural communities, often whispered instead of declared:

- "The big shaggy one"

- "Tree walker"

- "The Greyback"

- "Bushman"

- "The Black Giant"

One witness near Manitouwadge simply called it **"the other man"**.

None of these names suggest a cryptid. None suggest a creature in the sense we use the word. They describe something else. A presence. A peer. Something **intelligent**, but not human. Not part of our world—but very much in it.

Even across the U.S., this trend continues.

The **Skunk Ape** in Florida. The **Grassman** in Ohio. The **Woodbooger** of Appalachia. The **Momo** in Missouri. Some names are regional jokes. Others are rooted in fear. But they all describe the same general being—**tall, hair-covered, bipedal, fast, elusive, and uncatchable**.

And here's what matters: these stories didn't start in the internet age. They didn't start with YouTube or Facebook groups.

They've been around **for centuries**. Documented in letters, passed in oral tradition, mentioned in old trapper journals and 19th-century settler accounts.

These are not stories born from a single hoax or film.

They're **parallel truths** emerging in every forested region across a continent.

And the more you listen to what people *really* call it—the more you realize that "Bigfoot" is just a name **we gave something we didn't understand**.

The woods had names for it long before we ever showed up.

Part 2: What Language Reveals—Descriptions That Don't Know Each Other but Match

There's a strange kind of credibility that emerges when people, separated by thousands of miles and decades of time, describe the *same thing*—especially when they've never met, never read the same stories, and have no interest in fame or attention.

It's in those raw details, the *unpolished words*, that the truth starts to peek through.

I've interviewed people in Ontario who've never watched a Bigfoot documentary. People from traplines, logging families, and fly-in communities who describe encounters using **the exact same phrasing** as someone I spoke to in the Pacific Northwest—or in Louisiana—or in the mountains of Idaho.

They say things like:

- "It wasn't a bear."

- "It didn't run like a man."

- "The eyes weren't animal eyes."

- "It watched me like it was deciding something."

- "The smell hit me first."

- "It didn't want to hurt me, but it wanted me gone."

And they don't just use the same words. They react the same way.

Almost every credible witness goes through the same emotional arc:

1. **Dismissal** – They try to explain it away (bear, shadow, stump, sound).

2. **Realization** – They lock eyes or see movement and freeze.

3. **Disturbance** – Not just fear, but a **profound wrongness**—the sense of being watched by something that shouldn't be there.

4. **Silence** – Afterward, they don't tell anyone. Or they only tell one person. Sometimes not for years.

That's what gives these accounts their weight.

It's not that people are trying to match each other—it's that they *can't help* describing the same thing, because it's **what they actually experienced**.

Even more remarkable are the **descriptions of physical features** that repeat over and over:

- Wide, flat feet—sometimes with visible toe splay

- No visible neck—head set low between shoulders

- Arms longer than a human's—hands hanging below the thigh

- Hair covering most of the body—but not the face

- Skin on the face sometimes described as gray, or leathery

- A low, heavy breathing sound—even when the creature isn't moving

- A smell that comes *before* the sighting—described as musky, sour, wet-dog mixed with decay

And the eyes.

Always the eyes.

Not glowing, not like a cat's—but **reflective**, like moist glass. Usually red or amber in headlight reflection. Often described as "aware." Not animal. Not human. Just… aware.

These kinds of consistencies are hard to fake, especially when they come from people who don't want to be believed—they just want to be heard.

A man from Labrador once told me something that stuck with me ever since. He'd seen it once— near a cutline, in deep snow, while hunting alone.

He said: *"It wasn't like discovering something new. It was like remembering something I was supposed to already know."*

That's the feeling I've heard from others, too. Not terror. Not excitement. Just a quiet, rattled certainty that the world was never what they thought it was.

And here's what's terrifying:

None of these people knew each other.

None of them shared details.

And yet, they all *matched*.

In vocabulary.

In emotion.

In the eerie, unprompted sense that they had just encountered something that had been out there **all along**—watching.

Part 3: What the Old Names Hide—Cultural Memory and Regional Omissions

When people talk about Bigfoot, the conversation almost always centers on recent decades— sightings from the '70s forward, documentaries, digital footage, modern speculation. But the roots go **far deeper**, and much of what we think is new has simply been **renamed, reframed, or forgotten**.

There's a strange phenomenon in rural and Indigenous storytelling: **names are lost on purpose**.

Some cultures believe that speaking a thing's name too often gives it power. Others choose euphemisms to avoid drawing attention. You'll hear phrases like:

- "The tall ones"

- "The watchers"

- "The hairy people"

- "Them ones in the bush"

- "Old ones that walk at night"

In Northern Ontario, among Cree, Ojibwe, and Métis families, these soft-terms persist. They're rarely volunteered—you have to earn the trust to hear them. And when you do, they're often **accompanied by silences** longer than any sentence.

One elder I spoke with near Hearst told me, "They were always around. You didn't go to the marsh at night. You didn't whistle after dark. That wasn't for bears." When I asked what "they" meant, he only said, "You know what it means."

No flourish. No drama. Just a line that carried **generations of quiet understanding**.

And here's the key: these weren't considered mythical beings. They were seen as **co-existing presences**—kept out of conversation, but never denied. Their absence from modern names doesn't mean they weren't believed in. It means they were **too real** to be treated like fantasy.

In settler records, you find buried references to "wildmen," "shaggy Indians," and "upright apes," written off by trappers, traders, and missionaries as drunken tales or misidentifications. But if you look closely, the pattern is there:

- *Large barefoot prints found near creeks in early spring*

- *"Howling man-thing" reported near a Métis winter camp*

- *"Dark giant" seen stepping over a stone fence near Temagami in 1898*

These accounts were **filed away**, disconnected from any larger thread. No one at the time had the context to link them. But in hindsight, they match everything we've learned since.

Even regional folk tales hint at something more than metaphor.

There's the story of **The Sleeping Man of the Lake** told in hushed tones around Lake Nipissing. Or the **Shaking Forest** of Rainy River, where trees allegedly moved with no wind—and where people sometimes went missing for days, only to return silent and shaken.

Most historians chalk these up to weather or isolation-induced delusion. But when you cross-reference the locations with modern Bigfoot sightings, an eerie map begins to form.

One that spans centuries. One that transcends names.

And in that silence—between what was said and what was left unsaid—we start to glimpse the truth.

The real history of these beings didn't begin with the word "Bigfoot." That word is new. A label made to package something **old** in a form the public could digest.

But the truth?

The truth was already there.

Carved into stories.

Hidden in warnings.

Whispered between generations in a dozen languages we barely listened to—until the footprints showed up, and we realized we weren't discovering anything.

We were just **catching up**.

Chapter 17: Lakeside Shadows—Encounters Near Water

Part 1: Always at the Edge—Where Land Meets Silence

Ask any long-time wilderness traveler where the strangest moments happen, and they won't point to the dense forest. They'll point to the **shoreline**—especially the ones far from town, where boats barely go and roads vanish into moss and marsh.

There's something about water that draws both life and stories.

And it's at the edges—**where land meets lake**—that many of the most unnerving Bigfoot encounters have taken place.

It's not just because these are good places for food—fish, frogs, tender shoots—but because water muffles sound. Reflects light. Offers escape in every direction. And for something that avoids trails and cameras, **that matters**.

One of the strongest examples I've come across comes from **northeastern Ontario**, on a remote lake that I won't name here out of respect for the witness. He was staying in an old family camp, accessible only by bush road and canoe. Late one night, he heard splashing—not fish jumping, but **two distinct footfalls** through shallows.

He turned off his lantern and stepped out onto the deck.

From his description: *"Something was walking the shoreline, real slow. It was big. Upright. I could hear the water peel off its legs every step it took. Then it stopped—and I swear it was looking right at me. I didn't see eyes. Just this shape. Motionless. Like it was waiting for me to move first."*

He didn't go back to that camp for two years.

This story is one of many.

Water's edge sightings share a distinct pattern:

- The creature is **seen or heard first**, rarely both.

- They often occur at **dawn or dusk**, with still water and low wind.

- Witnesses report an eerie stillness—**birds go quiet**, even mosquitoes seem to vanish.

- And more often than not, the creature isn't moving toward water—but already **moving along it**, as if patrolling.

In several Northern Ontario cases, locals have found massive barefoot prints right at the lake's edge—**15 to 18 inches long**, with visible toe splay, spaced over four feet apart. The depth and spacing always suggest a heavy biped—but without dragging like a man would.

One of the more striking reports involved a family staying at a winterized cottage in late December 2013. They recorded video from their vehicle as they drove slowly down the snow-covered cottage road. There were **nearly a hundred prints** in a straight line. Each about **15 inches long and 8 inches wide**. You can hear the authenticity in their voices—the surprise, the awe. They eventually stopped and measured them.

The tracks ran parallel to the lake.

There were no ATV trails. No boot marks. Just clean, repeated impressions where something heavy and barefoot had walked along the snowy road—**right beside the shoreline**.

Some theorize this behavior—skirting the waterline—serves multiple purposes:

- **Muffling scent and sound**

- **Easy retreat into reeds or water**

- **Consistent game trails to follow**

- **Natural camouflage through fog and reflection**

Whatever the reason, the behavior is consistent.

And it's old.

Ojibwe oral stories refer to "the tall watchers by the water"—not river spirits or monsters, but walking presences that left wide, flat tracks and sometimes followed canoes from the shoreline without ever stepping into view.

In the end, it's always the same story: someone near water, hearing what they *shouldn't* hear.

Something stepping with intention, with weight, just out of sight.

And always, always staying just far enough away that **you start to question if you imagined it.**

Until you see the tracks the next morning.

And realize you didn't imagine anything.

Part 2: Reflection and Risk—Why Some Encounters Happen Near Camps and Boats

There's something about a lake that makes people let their guard down. Maybe it's the way sound travels differently—muted, almost dreamy. Maybe it's the illusion of space, how the open water gives you a false sense of control. Or maybe it's just the rhythm of being at camp: cast a line, sip a drink, leave your doors open, take your time.

And that's exactly when they show up.

Across northern Ontario and into parts of Quebec and Minnesota, reports of Bigfoot activity near lakefront camps share a disturbingly familiar set of details. Not sightings, always—but signs. Sounds. Movement. Sometimes, just a **feeling** that starts around dusk and doesn't go away until morning.

One report came from a couple staying at a cabin on a small fly-in-only lake northwest of Timmins. They were grilling fish on the deck when they heard something—*not in the woods behind them, but along the shore*, from the direction of the dock. They described a steady, slow splash. Not the flutter of minnows or frogs. **Heavy, deliberate steps in ankle-deep water.**

When they aimed a flashlight, nothing was there. But the footprints were—**wide, long, and spaced far too evenly to be human**. The next morning, they found an entire section of shoreline trampled, cattails snapped, mud churned up in a straight line along the water's edge. Nothing stolen. Nothing destroyed. Just... observed.

This kind of proximity is common around isolated lakes. In fact, it's **almost predictable**—if you stay quiet, don't run generators or blast music, and cook outdoors, something *might* drift closer than you think.

And it's not always at night.

I heard of a man solo-paddling a canoe on Lake Nipigon, following a meandering shoreline. As he rounded a bend, he saw a figure—tall, dark, crouched near the reeds. He thought it was a bear until it stood up on two legs and stepped silently back into the woods. No crashing. No run. Just a controlled exit.

He didn't talk about it for five years.

Boaters in other regions report glimpses of movement along the banks—shoulder-high shapes keeping pace for a few seconds before ducking behind trees. One Ontario angler even said he heard a **mimicked loon call**—perfect pitch, perfect rhythm—but with something "off." When he looked up, he swore he saw a head lower behind a stump just above the shoreline.

That's the unnerving part: the behavior isn't random. It's calculated. Precise. Measured not just in movement, but in **timing**.

Boats and lake camps offer windows. They isolate us. Sound travels. Light reflects. And when night falls, **we become the brightest thing on the lake**—easy to find, easy to monitor.

Campers often describe the same thing: something approaching just outside the lantern glow. A shape glimpsed between trees. Footprints in the damp earth near their canoes in the morning. Bait fish disturbed. Food wrappers moved.

But nothing ever stays long.

They don't linger. They don't loot. They come close, assess, and vanish.

You start to realize: they weren't curious.

They were verifying.

Confirming we were still as stupid, slow, and unaware as the last group.

Then they melt back into the bush.

Part 3: The Shoreline Stalker—Theories on Aquatic Proximity and Escape Routes

One of the more consistent features in regional Bigfoot sightings is how often the creature is spotted near water—but not **in** it. That distinction matters. It's rarely swimming, rarely fording a stream or splashing across a river. Instead, it's walking just inside the brush line. Skirting the edge. Following the shore.

That behavior raises a question: what's the strategy?

Some animals are drawn to water for food. Others for escape. But for Bigfoot, it appears to serve **multiple purposes.** The theory goes like this—**lakes and rivers aren't destinations; they're corridors.** Travel lines. Navigation aids. And most importantly—**paths of retreat.**

Let's break that down.

From an ecological perspective, water is a boundary. Most animals avoid staying near it too long unless they're feeding. It narrows their options. Forces them into the open. But Bigfoot seems to treat it differently. In dozens of credible reports, witnesses describe the creature **moving along the edge** of lakes, creeks, and marshes with ease. No panic. No hesitation. Just a silent, determined gait—like it knows the path already.

Take the case from **2013, south of Sudbury.** A family visiting their cottage recorded a long trackway—**nearly 100 prints**—that ran right along a snowy road near a frozen lake. There were no signs of disturbance in the surrounding woods, no weaving or erratic movement. Just a perfectly straight line. The prints measured **15 inches long and 8 inches wide**, each one evenly spaced, showing a stride far longer than that of a normal human. The family filmed it all as they slowly drove along, their voices full of genuine surprise and awe. At one point, they stopped and **measured the tracks by hand**. The result was clear: whatever made those impressions was moving with **deliberate, heavy steps,** and it had left behind one of the most compelling winter trackways ever captured on video.

That's not wandering.

That's **purposeful travel**.

And here's where the theory deepens.

Water not only offers camouflage—through mist, moonlight, and reflection—it also offers **escape routes.** Marshy banks. Beaver canals. Dense shoreline undergrowth. These are areas where humans can't follow without stumbling, making noise, or getting stuck. But if something is built for it—long legs, strong frame, wide feet—then those same hazards become **advantages**.

Think about the reports where the creature "just vanished." Now place those in the context of lakeside terrain. If something knows how to move through cattails without rustling them, if it can slip into a swampy inlet and vanish beneath a shadowed bank, then escape doesn't look like running.

It looks like **disappearing**.

Some researchers speculate that the creatures even use **old traplines and shoreline portage routes**—paths no longer mapped or used by people, but still there if you know where to look. These trails often follow natural water systems and curve along the backs of lakes that haven't seen human feet in decades.

It makes sense.

If you had to stay hidden, move silently, and always maintain an exit plan—you'd use the shoreline too.

Another piece of the puzzle: fish.

Several unconfirmed but compelling stories describe fish caches being raided near remote camps. Not bears. Not raccoons. The fish were **selected, not scattered**. Taken clean. Sometimes just the biggest ones. Sometimes the heads left behind. It's anecdotal, but suggestive. A few even report **rocks left behind** on docks or coolers—neatly stacked, like markers.

And here's where it gets eerier.

In multiple cases, **people report the creature doubling back** along a shoreline—looping around a small lake to approach a camp from a different angle. That's not instinct. That's **strategy**.

Why all this effort?

Because a shoreline can do something a forest can't—it gives you **sightlines**. The ability to see without being seen. Especially at night, when the forest is dark and the firelight is bright, when the canoe is pulled up, and the people are relaxed.

You think you're alone out there.

But the edges of water are full of eyes.

Some high in the trees.

And some **just above the surface**, standing still, watching the flicker of your fire from the reeds —**deciding**.

Chapter 18: When It Gets Too Close—Encounters Inside the Human Perimeter

Part 1: Lines Crossed—When Something Steps Too Close

Most Bigfoot encounters happen deep in the woods, far from towns and cabins. They're fleeting, often distant—seen crossing a trail, shadowed against trees, or heard walking away into the dark. But once in a while, something happens that breaks the pattern.

It doesn't just pass through the area.

It comes **closer**.

Into the yard.
Onto the porch.
Right up to the glass.

These aren't stories told over beers or pulled from fringe websites. These are **first-hand reports**, often delivered with shaking hands, low voices, and the kind of eye contact you don't fake when you've seen something you're not supposed to.

The pattern usually starts the same way.

A remote property. No neighbors for miles. An off-grid camp, a rural homestead, a quiet hunting cabin. Then come the sounds—subtle at first. Footsteps on gravel. Thuds near the tree line. A thump on the side of a shed with no wind. Tools left out are moved. A window screen gets torn.

Then it escalates.

One northern Ontario family near Wawa described hearing their garbage cans rattle around 2:00 a.m. Thinking it was a bear, the father went out with a flashlight. The can was upright. Nothing was missing. But across the driveway, pressed deep into the sand, was a **bare, wide footprint**—larger than his boot, with five clear toes.

Two nights later, they heard **heavy breathing** outside their window.

Another case came from a woman in Rainy River who lived alone on a small farm. She noticed things going missing—tools, firewood bundles, a bucket of potatoes stored in her porch cooler.

One night, her dog refused to go outside. She heard footsteps across her deck and a low grunt. The next morning, she found long smears of mud on her sliding glass door—**like something had leaned against it to look inside.**

She moved out within the month.

It's these kinds of stories—subtle, steady, escalating—that are the most disturbing. Not because of violence, but because of **intent**. The creature doesn't flee. It doesn't make a mistake. It *approaches*. Silently. Cautiously. As if testing the boundary between us and them.

And sometimes, that boundary is crossed.

In 2019, a family staying in a remote cabin south of Chapleau woke to loud bangs on the walls. The father grabbed a rifle and stepped outside. Nothing. But the **entire side of the building** had drag marks through the frost—**five long fingers traced horizontally across the aluminum siding**.

They left that day.

Other cases involve things peering through second-floor windows, standing behind woodpiles, or even being seen reflected in glass behind someone as they look out into the dark.

These aren't distant silhouettes or half-glimpsed shadows.

These are **face-to-face moments**—uninvited, unprovoked, and unforgettable.

And in almost every case, the witnesses describe the same feeling afterward: not fear. Not confusion.

Violation.

Like something intelligent had gotten just close enough to say, *I could come in if I wanted to. I just chose not to.*

And that's what keeps them awake long after it's gone.

Part 2: The Silence Test — Animals That Don't Bark and Generators That Won't Start

If you talk to enough people who've had close-range encounters — especially those involving rural homes or off-grid cabins — you'll begin to hear about something that doesn't show up in the photos or the footprint casts.

The silence.

It doesn't announce itself. It doesn't arrive with crashing branches or growls in the night. It settles in like fog. One minute the crickets are chirping, the wind's in the trees, your dog's shifting around on the deck. The next, **everything stops**. The sound drops out of the world.

No birds.
No frogs.
No insects.
Not even the hum of distant wind.

Witnesses say it's like the woods suddenly **hold their breath**. And in that silence, something changes.

Dogs — who moments earlier were restless or barking — go quiet. Some retreat inside. Some hide under decks. But more often than not, they just freeze. Ears up. Eyes fixed toward the treeline. As if listening for something the rest of us can't hear.

A man from near Red Lake told me he was out at his remote cabin when his dog suddenly backed away from the door — tail between its legs. He assumed it was a bear. But when he opened the cabin window to listen, **there was no sound at all.** Just stillness. A few seconds later, he heard **a single breath** — long, low, and exhaled close enough that it fogged the glass.

He didn't hear anything leave. But the dog wouldn't step outside until daylight.

There's more.

Generators that refuse to start. Trail cams that mysteriously fail — even brand new batteries drained cold. Flashlights flickering in unison. Radios crackling with static. These things could be written off as coincidence — until they happen **only** during the moments of presence. Right when the sounds go out. Right when the witnesses feel watched.

One Ontario man, staying alone at a fire tower outpost, reported his backup generator stalling each night around the same time—shortly after sundown. It worked fine during the day. But at night, right after the forest went silent, it would cut out. Then the tapping on the outer walls began. Slow. Deliberate. One knock. Then silence. Then two. Then nothing.

He left after the third night.

You hear similar reports from across the continent. Not from paranormal investigators. Not from ghost hunters. From **hunters**, **trappers**, **foresters**, and **off-grid workers**. People who don't scare easy. People who've heard wolves howl, bears growl, and cougars scream—and say this isn't that.

They talk about the silence like it's a **warning system**. As if the forest itself knows what's coming—and chooses to shut down until it's over.

Is it infrasound? Is it an effect of proximity? Or is it simply instinct—**a deep animal knowing** that something bigger, stronger, and smarter has arrived?

Whatever it is, it leaves a mark. Not a footprint. Not a scratch. But a feeling.

That something had entered your space, altered the rules of the night, and passed back into the dark—**unseen but not unnoticed**.

And in the end, that silence is what people remember.

Not the prints.

Not the smell.

Not the shape behind the trees.

The moment the world stopped moving.

The moment the forest went quiet—and something else moved in its place.

Part 3: One Thin Wall—Windows, Tents, and the Feeling of Being Studied

There's a specific kind of dread that only comes when you realize how little separates you from the outside. When the walls are thin. When the glass is cold and clear. When a tent flap is the only thing between you and a sound you can't explain.

It's not just fear—it's **proximity**. The deep, ancient knowledge that something is near and that you're exposed.

Some of the most unsettling Bigfoot reports don't involve sightings or physical signs. They involve **presence**—felt through tent walls, through windows, through the weight of stillness and breath.

A woman from Manitoulin Island described lying in bed one night at her family cabin, unable to sleep. The moon was bright enough to cast shadows through the curtain. Just as she began to drift off, she heard a *single step* outside. Then another. Then **breathing**—not panting or snorting, just slow, steady **inhalations**, right outside her window. She didn't move. She didn't speak. The curtain never lifted. But the breathing stayed for nearly a minute. Then it was gone.

In the morning, there were no tracks. No broken branches. But the screen had been **pushed in slightly**, like something had leaned close to look.

This kind of encounter happens more often than people think.

Tents are worse.

Something about being in fabric—no walls, no locks—leaves even the most seasoned outdoorsmen vulnerable. I've spoken to trappers, hunters, and solo hikers who admit they've **abandoned sites** mid-trip because of what happened outside their tents.

One man camping north of Kenora with two friends woke to footsteps circling their tent. At first, he assumed it was an animal. But it didn't sniff. Didn't scratch. It **walked. Upright.** Slowly. Deliberately. Around and around, pausing near the zippers.

They stayed frozen, whispering to each other in the dark, knives gripped in their hands. When it finally stopped, they didn't move for hours. Come dawn, they packed up in silence and paddled out.

Another account came from a group staying in a prospector's shack near Gogama. It was late autumn, snow barely forming, and they were inside playing cards. They heard **two knocks** on the far side of the shack. Then a *slow shift*—wood creaking, as if weight had pressed against the outer wall. One man moved to look through the window and froze.

He didn't describe what he saw. He just said: "It was standing there, and it knew I saw it. That was enough."

They kept the lanterns on all night.

The pattern in these encounters is unmistakable:

- **There's no attempt to break in.**

- **No noise meant to scare.**

- **No aggression.**

Just that feeling of being **studied**.

Not hunted. Not threatened.

Observed.

And in that quiet moment between breath and footstep, something ancient shifts inside you.

You realize the predator doesn't need to roar.

It only needs to watch.

From behind the trees.

From just past the flap.

From beyond the glass—so close you can feel the warmth of its breath, knowing that the only thing between you and it…

Is **one thin wall**.

Chapter 19: Burn Marks and Bluffs — When Bigfoot Sends a Message

Part 1: The Art of Intimidation — What Doesn't Happen by Accident

Most Bigfoot encounters involve avoidance. The creature keeps its distance, leaves signs but not itself, and watches instead of charges. But not all encounters follow that rule. Every once in a while, something **shifts** — a threshold is crossed, a perimeter violated — and what comes next is no longer passive.

It becomes **a warning**.

These moments often begin like the others: strange sounds, odd prints, a growing unease. But then the behavior becomes more focused — **targeted**. Not an accidental stumble into human presence. Not a curious brush with a trail cam. Something deliberate.

A family near Atikokan had been living in a hand-built log home, miles from the nearest town. They'd experienced the usual rural oddities: knocked-over trash, odd smells, the feeling of being watched. But one morning, they found a **burned ring** in their back field. No scorch marks in the grass — just a perfect circle where the soil had gone black and hardened, the surrounding ground untouched.

Later that week, they found **three trees bent toward the house** — not broken, just arched, held in place by woven branches. When they cut one free, they said it took hours to unbind without breaking. The arch was intentional.

Another couple, living year-round near Bancroft, returned home after a short winter trip to find **deep gouges across their garage door** — five long lines, like talons, each one evenly spaced and curved downward. They assumed vandalism at first. But their driveway had no tracks, no signs of vehicles. Just **those gouges**, and a faint muddy footprint below the door handle — **far too large for a man**, with no boot tread.

This isn't the work of a creature merely passing through.

These are **signals**.

There are dozens of reports like this across North America, with similar characteristics:

- Trees stacked or bent into unnatural shapes

- Rocks arranged in symbolic patterns—circles, lines, triangles

- Objects removed from homes and returned days later, rearranged

- Footprints placed **just outside windows**, as if to be found

One field researcher in Quebec described returning to a camp site where he'd left bait (not food, but brightly colored cloth strips and scent wicks). Nothing was taken. But a **log had been placed vertically in the center of his firepit**, standing upright in packed ash, surrounded by perfect bootless prints. "It felt like a challenge," he said. "Not an attack. But a line in the sand."

There's also the sound.

The **false charges**.

Many witnesses describe something rushing them—through the brush, snapping branches, sometimes bellowing. But just before impact, **it stops**. Dead silence. Nothing follows. No contact is made. When they retreat, the presence vanishes. This happens with hunters, hikers, even police officers.

It's not clumsy. It's **calculated**.

A warning that says: *I'm bigger. I'm faster. You won't see me coming next time.*

Even long-time researchers, those who don't believe in mysticism or telepathy or anything beyond physical biology, will admit: some behaviors feel **personal**. As if the creature isn't just avoiding danger—but responding to intrusion. Not animal instinct. **Territorial awareness**.

And in those moments, Bigfoot stops being a mystery hiding in the trees.

It becomes **a presence** making a decision.

To let you go.

Or not.

Part 2: Symbol or Instinct—What Broken Trees and Arranged Rocks Might Mean

In fieldwork across northern Ontario, one of the most common—and controversial—forms of evidence isn't tracks or audio. It's structures. Bends. Arrangements. Physical manipulations of the environment that appear too deliberate, too patterned, and too isolated to be natural.

Skeptics call it coincidence. Windfall. The work of time and gravity. And sometimes, that's true. But not always.

Because what researchers keep finding are **repetitions**. Patterns of **intention**. And these patterns often show up after an encounter—or in places where sightings cluster.

Let's start with **tree bends**.

In the boreal forests north of Thunder Bay and across the Algoma region, dozens of reports include mention of young trees—usually spruce, tamarack, or birch—**bent into arches** and pinned beneath logs or rocks. Not snapped. Not crushed. Bent slowly, deliberately, sometimes with no broken fibers in the trunk. These are not deadfall arrangements. They're targeted.

Some formations arch toward trails. Others face water. A few are woven with secondary branches pulled from other saplings. You don't see these deep in untouched forest. You find them **on the edges**—where people have been, or might return.

Then there are **rock stacks**.

Balanced stones, in threes or fives, found along rivers and on game trails. Some are clearly human. But others? Found miles from the nearest access point, in terrain requiring full-pack hikes and no evidence of prior camp. The formations are often low to the ground—stacked carefully, with the largest on top. Not for aesthetics. For **balance**.

One Ontario researcher found a ring of rocks surrounding a half-buried footprint on Crown land. He hadn't announced his survey. He hadn't baited the area. But the print—and the ring—was fresh. Not ceremonial. Not for art. Just... placed.

So what does it mean?

Are these **symbols**, meant to communicate? Or are they **instinctual behaviors**—territorial markers like a wolf's scrape or a deer's rub?

The answer might be both.

In primates, we've seen tool use evolve into **ritual behaviors**. Chimpanzees stacking rocks in trees, not for any immediate gain, but seemingly to mark presence. Gorillas creating **drape shelters** of broken foliage to sleep beneath. Elephants rearranging logs and bones. Why wouldn't a creature with upright posture, opposable thumbs, and apparent pattern recognition do the same?

Especially if it **avoids vocalization**. Especially if it prefers **visual communication**—silent, unambiguous, durable.

Here's where it gets unnerving: Some of these signs appear **after confrontation**. After property damage. After sightings where witnesses stood their ground. Tree arches found days later, facing cabins. Logs placed across driveways that had been cleared the day before. A child's toy taken from a porch and returned—**stacked carefully between stones**.

Not random. Not ritualistic. Intentional.

Like leaving a calling card.

Not as a threat. But as a way of saying: *We were here. We saw you. Don't come back.*

To some, this sounds like imagination. But to the witnesses—especially the ones who live off-grid, who know every inch of their land—it's all too real. They're the ones who find **structures where nothing stood the week before**. They're the ones who hear **knocks after disturbing an arch**. They're the ones who stop clearing new trails—not because they're afraid, but because they finally **got the message**.

This isn't a creature marking food.
It's marking **territory**.
And maybe, just maybe, **trying to make itself understood**.

Part 3: The Edge of Aggression—Bluff Charges, Threat Displays, and What Might Come Next

There's a fine line between intimidation and attack. Most wildlife encounters fall on one side or the other. A black bear huffs, stomps, maybe false-charges, then retreats. A moose lowers its head, signaling you to back off. But when it comes to Bigfoot reports, something sits right on that edge—where the behaviors are neither full retreat nor open confrontation.

They are **displays**. Threats not carried out, but unmistakably **personal**.

Take the 2020 incident from a family just south of Hearst. They'd been hearing strange knocks and movement in the woods for two nights. On the third, while sitting around a fire, they heard a **long, chest-deep growl** from no more than 40 feet into the dark. Not a bear. Not a coyote. Something larger. The growl rose to a **shuddering vocalization**—a kind of roar that made their chests vibrate. Then came the **charge**—a crashing sprint that stopped just short of the firelight. No figure seen. Just pressure. Presence.

Nothing followed. No attack. Just a deliberate approach and sudden stillness. Then, minutes later, the sound of **branches snapping**—not randomly, but in sequence, as if whatever it was wanted them to hear its **exit**.

These bluff charges, as they're sometimes called, come up again and again. And they almost always follow a **pattern**:

1. A knock or vocalization.

2. Stillness.

3. A sudden crash or movement—fast, violent, but never completing the approach.

4. A retreat—loud or quiet, but clear.

A man near Temagami described it like this: "It wanted me to know it could break the rules. That I was lucky it didn't."

Another common thread in these encounters is **object displacement**. Logs hurled. Rocks thrown —not just dropped from above, but flung with enough velocity to strike trees, tents, even vehicles. In several cases, the objects land **near** the witness, but never **on** them. Close enough to startle. Not to wound.

Is it restraint? Or precision?

A particularly memorable case I investigated took place just off **Highway 6, north of Manitoulin Island**. Two smelt fishermen were working the riverbank after dark, **scooping with nets** as the smelt run tends to peak late into the night. They stood side by side—one wearing a headlamp, the other wielding a long-handled net—**moving quietly through the shallows**, dipping and lifting with practiced rhythm, the only sounds the shuffle of boots and the swish of water.

The night was still. No traffic. No human sounds. Just the crackle of current and the flicker of light across water. Then came a **sudden crack**, followed by a **massive splash**—not the kind a fish makes, but something heavy—**a basketball-sized boulder**—striking the river with force. The water erupted within feet of them, soaking their boots and sending them stumbling back from the edge.

The rock had come from across the river, launched with incredible power from the dense treeline. They swept the area with flashlights but saw nothing. No movement. No silhouettes. No fleeing animal. The bush on the far bank was thick. Inaccessible. And silent.

One of the men said to me afterward, "That wasn't a slide or a roll. That was **aimed. Thrown.** I was scooping with the net—barely making noise—and it hit the water right next to me. Whatever did that, it was watching."

There was no logical explanation. No cliffs above. No machines. And no human seen or heard. The boulder was far too heavy and precise for coincidence. What struck them hardest wasn't the rock—it was the **message**.

It matched other reports I've taken across Ontario: the **targeted throwing of heavy objects**, not in rage, but in **controlled intimidation**. No follow-up. No bluff charge. No scream. Just one timed act—like a warning.

A line in the dark that said: *I know you're here. I could come closer. But I don't need to.*

And here's what makes it chilling: most of these encounters happen **after a human has pushed too far**—ventured off trail, crossed a creek bed, followed tracks too long. As if the creature had issued silent warnings, and the person kept going.

Until it decided to **respond**.

In this way, the behavior mirrors **territorial defense** found in apex predators—but with something more complex. Not just posturing. **Communication**. A calculated show of dominance that stops just shy of contact.

Because in every one of these encounters, the creature could've done more.It could've stepped into the light.It could've closed the distance.It could've broken the glass.

But it didn't.

And that's what makes the silence more unnerving than any roar. It says: *We're aware of each other now. You've been warned.*

What comes next?

No one really knows. But seasoned researchers will tell you: if something bluff-charges you, throws something, or sends a sound that hits you in the chest like a subwoofer—it's not curious.

It's drawing a line.

And it's your move next.

Chapter 20: Body Count—What About the Missing?

Part 1: Where the Trail Goes Cold

For all the sightings, sounds, tracks, and theories, there's a darker corner to this subject—one that even many field researchers avoid. It's the part that's harder to prove, impossible to document, and full of uneasy speculation.

The **missing**.

North America's wilderness eats people. That's a fact. Between hunters, hikers, foragers, prospectors, and campers, hundreds disappear each year. Most are eventually found—injured, lost, passed away from exposure. But some... some are never located at all. No clothing. No gear. No trace.

Sometimes, the circumstances surrounding those disappearances push beyond what can be written off as accident.

Take the case of a man in his 60s, a seasoned bush pilot who disappeared in 2015 after landing near a lake west of Chapleau. His floatplane was found intact, beached exactly where it should have been. His gear was still inside. A firepit had been set up. His boots were dry. No signs of struggle, no animal activity, no distress. And no body.

The OPP called it a wilderness misadventure. But several other locals quietly mentioned strange lights seen around that lake for years. Knocking sounds. A howling that wasn't wolf or loon.

There are clusters like this across Ontario—unsolved missing persons cases where the terrain didn't match the outcome. Where search dogs refused to track. Where helicopters flew grid after grid and found nothing. In some cases, tracks led up a slope... and stopped.

One hunter near the Temagami region went missing less than a mile from his truck. His GPS said he never left the trail. But his backpack was found hanging in a tree, 11 feet off the ground. His rifle was wedged into a split in the trunk, upright. Like it had been placed there.

He was never recovered.

This isn't about suggesting Bigfoot is abducting people. There's no evidence of that. But when you study the subject seriously, these disappearances raise **questions**—not about monstrous behavior, but about proximity. About what happens when something wild gets too close… and doesn't care to be seen.

There are also the stories that never make it to police reports. The ones whispered by old hunters. The warnings you hear from bush workers to "stay on the logging road after dark." Accounts of **something watching**, following, sometimes even **shadowing camps** for days.

It's always the same message: **"Don't go alone."**

One retired forestry tech told me flat-out, "I've seen black bears, wolves, cats, and drunk men with guns. None of them ever made me leave camp early. But this? Whatever it is… you feel it first. Then you hear it. Then it doesn't leave."

He wouldn't say what "it" was. Just that it didn't belong, and it knew he knew.

So how many disappearances can be explained? Probably most. But in the far margins—where gear is found folded, where boots are dry, where the last radio call is just static and the dogs won't cross the treeline—those are the cases that sit differently.

Because if you believe Bigfoot is real—an undiscovered, highly intelligent primate adapted to avoid contact—then you must also admit it's capable of **territorial defense**, **evasion**, and maybe even **choosing confrontation** in rare cases.

And that raises the quietest question of all.

What happens when someone gets too close… and doesn't leave?

Part 2: Evidence vs. Theory—What We Know and What We Fear

If you talk to professionals—search and rescue veterans, retired game wardens, experienced hunters—they'll all tell you the same thing: people go missing in the bush. Terrain changes. Weather turns. People panic. It's tragic, but it happens.

But then there are the outliers. The **missing cases** that just don't line up with how the wild usually works.

There's a term for it in some circles: **"Silent Vanishings."** These are the cases that involve:

- No distress call

- No signs of a struggle

- Gear left behind in strange ways

- Footprints that just stop

- Search dogs that act confused, disinterested, or outright frightened

These aren't the stories of someone falling into a ravine or getting lost after a sprained ankle. These are events where seasoned outdoorsmen—people who know how to navigate, who had communication tools, and often had backup—simply **disappear**.

And the theories start to split into two lanes.

The first is **environmental oddity**. Unmapped sinkholes. Sudden weather systems. Predator attacks that leave no trace. Unlikely, but technically possible.

The second is where it gets darker: **deliberate, intelligent interference**. Not the paranormal, not interdimensional doorways or cloaking fields, but the theory that **something was aware of the person**—and took action to remove them, silence them, or send a message.

This is where Bigfoot enters the conversation—not as a cryptid lurking at the edge of myth, but as a **flesh-and-blood creature with a psychological perimeter**, a sense of territory, and maybe even a threshold for intrusion.

It's rare. It's not what the majority of sightings show. But it's in that rare percentage—the last half-percent—where encounters grow colder. Where the witnesses never make the report.

Take the **trapper from near White River** who vanished in 2018. His camp was found in perfect order. A kettle over the firepit. Traps still laid out on a line. His parka was hanging on a nail inside the cabin. No signs of animal entry. But a clear human-sized track—barefoot, in snow— was found **just behind the cabin**, leading toward the woods. One print. That's all. Not his boot size. Not his weight. Not human.

Police dismissed it. Locals did not.

Researchers are often asked why there's no body. Why no remains have ever been conclusively linked to Bigfoot. But ask the same of mountain lions. Of bears. Bodies in the wild vanish fast.

Within days, sometimes hours, carrion animals erase evidence. Rain collapses tracks. The forest **absorbs what doesn't belong**.

And maybe that's the scariest part.
Not that people disappear.
But that they do so **cleanly**.
Without explanation. Without closure.
Like the forest reached out and **edited them out** of the narrative.

It's not proof of anything. But it's a pattern.
And when too many patterns start to overlap—sightings, sounds, structures, intimidation displays, and disappearances—you're no longer talking about coincidence.

You're talking about something **with rules**.
Rules we don't understand.
Rules we probably break every time we step too far off trail.

Chapter 21: Global Kin—Other Creatures, Other Names

Part 1: It's Not Just Here

North America may be where the term *Bigfoot* was born, but the creature itself—at least, something remarkably like it—has been reported across nearly every continent for hundreds, even thousands of years. The patterns aren't isolated. The names change, the terrain shifts, but the stories echo each other with eerie consistency.

Start with **Russia**—the snow-covered wilds of the Caucasus, the Ural Mountains, and remote Siberia. There, the creature is known as the **Almasty**—described as upright, muscular, often shy but sometimes aggressive, with hair ranging from gray to brown. Local reports speak of it stealing livestock, peering into windows, and sometimes walking right into villages. Russian researchers treat it seriously. Expeditions are conducted. Casts are taken. In some circles, it's not even called a cryptid. It's considered **a species that's evaded formal classification**.

In the **Himalayas**, there's the **Yeti**—larger than life in Western lore, but described by locals as smaller than pop culture suggests. Stockier. White-furred only in winter. And always—always—seen in the high passes where few dare go alone. Monks and mountain people alike speak of it in reverent tones. Not as a monster. As something wild and aware.

China has its **Yeren**, most often reported in the Shennongjia forests. Descriptions match North American Bigfoot almost to the inch—broad shoulders, deep-set eyes, long arms, no tail. Government scientists have conducted formal studies. Dozens of footprints have been cast. And while no physical specimen has emerged, enough credible accounts exist to keep the search active.

Move to **Australia**, and you'll hear of the **Yowie**. Aboriginal cultures have spoken of it for generations, and white settlers began recording sightings in the early 1800s. It's often linked to water—rivers, billabongs, and coastal forests. Yowie behavior ranges from passive observation to full intimidation displays: rock throwing, growling, even rushing tents. Familiar?

In **South America**, particularly Brazil and the Amazon region, there are stories of the **Mapinguari**—a tall, foul-smelling creature with backward-facing feet and a piercing vocalization. Some versions may blend folklore with prehistoric memory, but locals report modern encounters with something big, bipedal, and territorial in the forest margins.

Even parts of **Europe** have their candidates. The **Wildman of the Pyrenees**, the **Skogsrå** of Sweden, and ancient Celtic and Basque stories all point to upright, hairy, forest-dwelling beings with humanlike intelligence and inhuman strength.

And these aren't just ghost stories.

They often come with:

- **Footprint evidence**

- **Consistent anatomical details**

- **Similar vocalizations**

- **Behaviors that mirror those seen in North America**

So what does it mean?

It suggests that whatever Bigfoot is, it's not a one-off. It's not the product of a single culture's imagination or a North American curiosity. The core elements—upright posture, stealth, massive size, territorial behavior, vocal mimicry—appear in **unconnected, pre-technological cultures**, often in areas where human intrusion was rare.

This isn't global hysteria. It's **species persistence across oral tradition**.

Which brings us to the real question: if so many cultures describe the same creature, is it a case of **cultural overlap**, or are we looking at a **globally adapted relic species**?

If it's the latter, Bigfoot isn't just North America's secret.

It might be the planet's best-kept one.

Part 2: Language of the Wild—Shared Traits in Sightings Worldwide

When you strip away the cultural varnish—the local names, mythologies, and artistic interpretations—a clear pattern begins to surface. Around the world, from high-altitude plateaus to jungle basins and coniferous forests, people describe creatures that share a common behavioral and physical template. It's as if they speak a **common biological language**, whether we call them Bigfoot, Yeti, Almasty, Yowie, or something else entirely.

Let's look at the **shared traits**, across geography, belief systems, and ecosystems:

1. Bipedal Movement
In almost every credible account—from the Appalachian backwoods to Nepalese highlands— these creatures are described as walking upright, with a **natural, balanced gait**. Not hunched like a bear or clumsy like a gorilla forced onto two feet. Witnesses consistently describe long strides, fluid motion, and an ability to traverse rough terrain with **ease and speed**.

2. Massive Size and Musculature
Descriptions range from six to ten feet in height, but always with **unusual bulk**—broad shoulders, thick torsos, and powerful limbs. In some Yeren and Yeti accounts, the creatures are described as "barrel-chested" or "like a wrestler made of fur." This isn't folklore inflation—it's **consistent anatomical scaling**, even in populations separated by language and technology.

3. Stealth and Evasion
Despite their size, these beings are rarely caught in the open. They move silently, avoid open areas, and seem to have an uncanny sense of **when they're being watched**. From the Mapinguari to the Yowie, reports mention creatures "disappearing like smoke," "fading into trees," or "melting into the bush." Some researchers theorize this points not to paranormal ability, but to **an evolved mastery of terrain and cover**—possibly even an innate understanding of human sensory blind spots.

4. Rock and Tree Interaction
Whether it's **thrown stones in Australia**, **tree arches in Siberia**, or **stacked branches in Ontario**, the interaction with natural materials appears worldwide. Many of these behaviors go beyond instinct—they feel communicative. Or at least **territorial**. It suggests a cognitive level above simple reaction. A choice to **warn, intimidate, or mark boundaries**.

5. Vocalization
Screams, whoops, howls, chest beats, deep resonant "booms"—these sounds are reported globally. They don't match known animal calls and often carry **tremendous volume and emotional impact**. In some cases, witnesses report feeling the sound more than hearing it— **chest-pressure, inner-ear vibration**, and even disorientation. This is remarkably similar across continents.

6. Odor and Presence

A strong, musky, almost rotten smell is reported in many encounters. Descriptions include "wet dog," "rotting cabbage," and "sulphur mixed with sweat." These odors often appear before any visual contact, suggesting a **chemical marker**, not unlike territorial scenting in known species. It's also theorized to be a defense mechanism—a primal warning that says: *Leave now*.

7. Avoidance of Technology

A subtle but persistent thread in sightings is the **lack of photographic or video evidence**, even in areas under surveillance. Trail cams, drones, phones—all fail to capture meaningful images. This phenomenon isn't just in North America. Yowie researchers in Queensland, Almasty trackers in the Caucasus, and Yeti seekers in Nepal all report the same strange absence: **evidence just beyond the reach of the lens**. Whether this is luck, awareness of camera placement, or simply a byproduct of low encounter rates, the pattern persists.

8. Water Proximity

Reports from Brazil to British Columbia often note rivers, lakes, and marshes as the **common ground**. Whether for hydration, food sourcing, or mobility, these creatures seem to use water as a **natural corridor**. Even the Mapinguari, often thought of as jungle-based, is frequently seen near flooded forest edges and swamp systems.

What's more compelling than these traits individually is how often they **overlap**—despite being separated by oceans, culture, and generations. No central story passed them between continents. Yet people from vastly different backgrounds are describing **the same creature**, with uncanny detail.

This doesn't prove existence.

But it does prove consistency.

And when the unknown keeps repeating itself in the same voice, across the world's wildernesses, maybe it's time to stop calling it a myth...

...and start calling it **a pattern**.

Chapter 22: The Ones Who Don't Talk—Hunters, Rangers, and What They Really Know

Part 1: Off the Record

You won't find their stories in newspapers. They won't post about it on forums. But they exist—seasoned hunters, provincial rangers, retired law enforcement, wildfire lookouts, and trappers who've spent **decades in the bush**. Men and women who live closer to the land than most people drive to work. People who know the difference between a moose breaking brush and something else staying just out of view.

And they almost never speak publicly.

Sometimes it's job protection. Sometimes pride. More often it's something simpler: they don't want to be laughed at. Or worse—**believed**.

I've sat with more than a few of these people. Some on porches in remote Ontario towns. Some standing over maps with a beer in one hand and a finger tracing river routes. One or two have just mailed in photocopies—**no return address, no names**.

What they tell me is always the same:

"I know what I saw. I just don't talk about it."

One forestry worker near Hornepayne told me he'd heard **"a voice in the trees"**—not English, not a language at all, but something rhythmic, pulsing, spaced like speech. It echoed for minutes, then went silent the moment he moved toward it. He didn't report it. He just packed up and left.

A former game warden described seeing **a large figure kneeling in cattails**, silhouetted by moonlight, fishing something from the shallows with its bare hands. When it stood, it turned toward him. Its eyes caught the light. He said it looked at him like **a man deciding whether to speak**—but then just stepped backward and was gone. He told no one at the time. Not even his family.

And there are more.

The fire lookout operator who saw something upright cross a burn zone in one smooth motion, then vanish into black spruce.

The bear hunter who followed a trail of gut piles from a bait site that led to a single massive track.

The trapper who kept finding **his sets sprung, but untouched**, with drag marks nearby—like something **lifted the steel jaws clean off the ground and set them down.**

None of them want credit. None of them are looking for fame. They just want the silence between the trees to make sense. Or at least to be acknowledged.

In a way, these are the most credible witnesses. They have **nothing to gain**, and often something to lose. Their reputations. Their standing in tight-knit rural communities. Their trust in the job.

But occasionally—after dark, with no recorder running—they'll talk.

And they'll use a phrase I've come to respect more than any book, podcast, or database:

"You don't have to believe me. Just don't follow it."

Part 2: Cabin Logs, Helicopter Crews, and What Gets Left Out of Reports

Some of the most compelling stories I've heard never made it into official channels. They live in **cabin logs, radio chatter, and the pauses in conversation** when someone knows exactly what you're asking but chooses not to answer directly.

Take the case of a **fire tower operator east of Cochrane**. For nearly 15 years, he logged every anomaly—weather, lightning strikes, illegal burns, even the occasional bear encounter near his post. But in one stretch of summer 2014, his entries changed. The language got vaguer. He wrote things like:

"Unidentified movement below SW quadrant, 11:40 p.m."
"Light interference again. No source found. Not aircraft."
"Nocturnal vocalizations continue. Unknown origin."

That logbook, passed to a contact of mine, showed **twelve nights of recurring activity**. Each entry danced around the same theme—something big, something bipedal, circling the perimeter of his range, just out of sight. When I reached out to him later, he said only:

"They told me not to speculate. So I didn't."

Then he added, "But I didn't sleep right for a month."

Wildland firefighting crews in Northern Ontario have told similar tales. One helicopter pilot described **a figure walking along a ridge near a remote burn zone**, moving too smoothly for a man in gear and too large for a deer. They circled back—gone. Not hidden. Gone.

Another former MNR employee recalled night sweeps for injured moose during collaring seasons. One night they saw **eye shine—red, steady, about nine feet high.** They thought it was a reflector from an old logger's marker. But when they moved closer, it blinked... and vanished. The next morning, fresh tracks were found. **No known animal matched the depth or gait.**

None of this was submitted in official reports.

Why? Because most agencies don't have a form for it. And those that do—unofficial logs, internal memos—treat such reports like a campfire story: notable, maybe, but irrelevant to policy.

There are even rumors that some reports are quietly **redirected or buried**. I can't confirm that. But I can confirm this: I've met rangers and bush pilots with **photographs they won't release**. Not because they're afraid of disbelief, but because they fear exactly the opposite.

"What if it's real?" one of them asked me.

"What if I release this, and they come looking for it?"

"What if they come looking for me?"

He wasn't being dramatic. He was being honest.

You see, many of the people who live and work closest to the land aren't afraid of ridicule. They're afraid of **breaking an unspoken contract**: respect the forest, and it leaves you alone.

They've seen the dark shapes.

They've heard the knocks, the calls, the breathing outside their tents.

And they've chosen silence not out of fear... but out of something older. Something like **respect**.

And then there was the story shared by a **retired First Nations police chief** during a local northern Ontario podcast—not a sensational one, just a quiet, community-based show where guests usually talk about moose hunting, ice roads, or snowmobile trails. But this story was different. His tone was careful. Reflective. Like he still wasn't sure he should be saying it out loud.

He had been flying from **North Bay to Sudbury**, seated on the right side of a charter plane. As they descended for approach into Sudbury airspace, the aircraft banked low over a stretch of **deep bush—no logging roads, no signs of development**.

There, in a long, snow-covered clearing, he spotted something he wasn't prepared for. A **massive, upright figure**. Moving at a steady pace. Its arms swung low and wide. It wasn't clumsy, like a man slogging through snow. It was **deliberate, smooth**, and clearly alone.

He still flies. Still looks out the window. But now he watches the tree lines with a different kind of attention—**not looking for landmarks, but for movement. For that one moment where the forest forgets to hide what lives inside it.**

Chapter 23: Glimpses in the Frame—The Best Video and Audio Evidence to Date

Part 1: The Frames That Refuse to Die

For over half a century, people have been pointing cameras at forests hoping to catch proof of what shouldn't exist. Most attempts end in frustration: empty clearings, blurred shadows, strange shapes that could be anything. But now and then, something surfaces that refuses to be dismissed —**images and sounds that stick in the mind like a splinter**.

It always starts with the most famous of them all: the **Patterson-Gimlin Film**.

Shot in Bluff Creek, California, in October of 1967, the short clip shows a large, upright creature striding through a dry creek bed. The figure—now known as *Patty*—turns its head mid-stride in a movement that feels too smooth for a suit and too casual for a hoax. The footage has been stabilized, enhanced, dissected, and debated for decades. Skeptics call it a man in a suit. Supporters cite the muscle movement, gait mechanics, and limb proportion as nearly impossible to replicate in 1960s costume tech.

The real story isn't just the film itself. It's that **no one has replicated it convincingly**, even with today's technology. Every remake lacks something—weight, posture, anatomical detail, or simply that strange, haunting fluidity.

But it's not the only video.

In 1994, a hiker in the Sierra Nevada captured what became known as the "**Blue Mountains footage**"—a tall, dark figure descending a slope, walking in a biomechanically correct manner, then disappearing into thick timber. Analysts noted the **shoulder roll** and **weight shift**, which matched primate motion patterns.

Then there's the **Marble Mountain footage**, filmed in 2001 by a youth group in Northern California. It's long-range, shaky, and far from conclusive—but the figure's sheer scale, visible even from hundreds of feet away, continues to baffle scale-based analysis.

In Canada, the **Mission, British Columbia trail cam image**—captured in 2012—shows a bulky figure mid-step in a forest corridor. It was dismissed by some as a man in dark gear, but there was one problem: **no trail access** where it stood. No boot prints. And the next five frames? Empty. Like it knew when to step off camera.

But perhaps more convincing than any photo are the **sounds**.

Recordings from Ohio, Washington, and Northern Ontario feature **deep howls, whoops, and knocks** that have stumped biologists. In 2015, a team in the Adirondacks captured a 20-second howl that traveled over two ridges—**not wolf, not coyote, not moose.** The spectrum analysis showed **infrasonic components**—low-frequency tones outside human hearing range, often associated with elephants and whales.

Several sound engineers have examined these clips and concluded one thing: **they aren't edited.** The vocal tracts required to produce these tones would need to be massive. Not impossible. Just not human.

One of the most unnerving pieces came from northern Michigan: a clip known as the "**Samurai Chatter.**" It features a strange, rapid vocalization captured on a parabolic mic—guttural, tonal, almost like language. No known animal speaks this way. Some compare it to **proto-speech**, or the muttering of someone talking to themselves in a dialect no one else knows.

We have clips. We have frames. We have audio.

What we still don't have is **closure**.

Each new video is chased by grainy doubt. Each sound dissected by forums. The clearer the evidence becomes, the more people seem to **step away from belief**, not toward it—afraid, perhaps, that acknowledging the real would change the world they think they understand.

But these fragments persist.

Little windows into the unknown.
Unclaimed truths caught between frames.

Part 2: Ontario's Untold Clips—Footage, Sound, and the Ones That Went Missing

Ontario is home to vast, unmapped wilds, some of the most **lightly traveled boreal regions** on the continent. And while it's rarely front and center in the media, there is no shortage of visual and audio evidence—**most of it never makes the rounds online**. Some of it is lost. Some is withheld. And some simply gets ignored.

A man I met in Hearst showed me a **grainy cell phone clip** taken during a moose hunt near the Nagagami River. The video is unremarkable until halfway through—just wind, spruce, and the shaky breathing of the man filming. Then a deep, **wood-on-wood crack** slices through the background. Followed by a second. Then silence. You can hear the man whisper:

"It's back. That's the same thing from last week."
He turns the phone to scan the treeline—nothing but forest. But the knocks continue, **rhythmic and pulsing**, as if someone were trying to draw attention… or issue a warning.

He never uploaded it. Didn't want the attention.

"I'm not a YouTube guy," he told me. "I just want to know if anyone else heard it."

Another case came from just south of Temagami, where a **fisherman testing a trail cam** accidentally recorded what he first thought was a bear. The footage—now corrupted after a botched SD card recovery—showed a **broad, dark shape passing low** through the field of view. But when he scrubbed the video frame by frame, he caught a clear image: **a knee bending backwards**, like a man crouching, but far bulkier, more animal than human.

He sent it to a family friend in the MNR who offered to have it analyzed.
He never saw it again.

"I think it spooked him," the man said. "He said, 'you don't want this shared.' Then he went quiet."

But the most haunting clip I've encountered from Ontario never made it to video.

It was **audio**, recorded just outside of Chapleau in 2020 by a trapper who was testing new audio monitoring equipment. He'd set it up to track wolf activity across a frozen corridor near his cabin. What he captured instead was **a sustained, low moaning sound**, somewhere between a scream and a foghorn—deep, layered, and impossible to localize. It lasted twelve seconds. Then silence. Then the sound of something **large crashing through thick timber**.

I heard it once. Just once.
The trapper let me listen through old headphones in the back of his snowmobile shed. When I asked why he never shared it, he looked me in the eye and said:

"You think people want answers? They don't. They want to be entertained."

I offered to help him release it anonymously. He shook his head.

"You do that, and I lose this place."

So instead, the clip lives on a dusty thumb drive wrapped in aluminum foil and stuffed into the back of a waterproof case next to spare batteries and an old GPS.

That's how fragile this evidence can be. **It exists. But just barely.**

Most people will never see these frames. Never hear these calls. Not because they're fake—but because the people who hold them are **more afraid of attention than disbelief.**

And maybe that's what makes them worth listening to in the first place.

There's also the **Dryden case**—a family out for a casual outing in the woods who managed to capture **some of the most chilling vocalizations** I've ever heard from Ontario. The video is short, handheld, and unpolished—but the sounds it contains are unforgettable. Raw. Piercing. Haunting. I believe it's the **best ever captured in this province. They made it public, so it didn't vanish.**

Chapter 24: Not a Bear — Debunking the Common Dismissals

Part 1: The Wrong Animal, the Wrong Behavior

The first thing anyone says when they hear a Bigfoot story is the same, almost word for word:

"Probably a bear."

It's a reflex. A defense. A placeholder for uncertainty. And on the surface, it makes sense. Bears are big, they stand upright, they're elusive, and they live in the exact habitats where most sightings occur. But as any serious field researcher will tell you, **there's a big difference between a bear and what we're describing**.

Let's start with posture. A bear can rise on two legs, but it doesn't stay there long. It's a behavior reserved for **scanning, bluffing, or sniffing**—not travel. When it does walk upright, it's clumsy, waddling, off-balance. The knees don't bend like a human's, and the stride length is short, awkward. Bears don't cover distance efficiently on two legs. But witness after witness describes **fluid, long-strided motion—four to five feet per step**, arms swinging, head turning mid-stride. Not only upright, but **naturally upright**, like it was built for it.

Then there's the matter of anatomy. People who claim to see Bigfoot often describe **wide shoulders, no visible neck, long arms hanging low near the knees**. The hands—if seen—are described as wide, with fingers. The head is sometimes conical, the back sloped. None of this matches a bear's silhouette.

And the feet.

Black bears leave distinct tracks: **five toes, inward-angled heel, claw marks** in front of each digit. But the tracks attributed to Sasquatch—seen over and over in snow, mud, and clay—show **flat, wide impressions**, no arch, no claws, with **a pronounced midtarsal break**. That's something you see in non-human primates, not bears. And it's something the average witness wouldn't think to fake.

Behavior is the final piece.

No one mistakes a bear for watching them from behind a tree. No one says a bear threw a rock. No one reports **parallel pacing through the brush for twenty minutes** without a single sound. But these are common in Sasquatch encounter reports. Repetition. Silent observation. Sudden withdrawal. Then nothing.

I've spoken to seasoned hunters—men who know every bird, every track, every sound in the bush—who thought they were being followed, only to turn and find **nothing there. Just presence. Just pressure.** And then a tree knock. Or a whistle. Or a rock clacking against another.

One hunter from just west of Sault Ste. Marie told me this:

"I've walked into bears more times than I can count. You feel it. You smell it. You hear the huff, the stomp, the crash if it bolts. But this? This was different. I felt it first. Like being watched before you even know where to look. Then it moved behind me without making a sound. That's not a bear. Bears don't do that."

But I also understand where the instinct to blame a bear comes from. Because I've seen what a bear can do, and I've seen how fast your brain wants to explain what it's witnessing.

Years ago, just outside of Sudbury, I had an incident that burned itself into my memory.

My son and I were walking a bush trail with my German shepherd, Chase. It was late in the day, quiet and overcast. My son was about thirty feet ahead, and I called out to him:

"Stay close. You don't know what's around the bend."

Just as I said it, he froze.

Chase moved forward slowly, ears up, focused, no barking. He reached the bend and stopped. That's when I saw them—**two bear cubs racing up a tree**, clearly startled. Chase quietly moved between them and my son, placing himself in the exact center of the tension without making a sound.

I got my son behind me and told him to back up slowly.

Then I heard it—**clicking noises**, twenty feet ahead, just around the curve of the trail. Sharp, deliberate, not like twigs or claws, but like something tapping rhythmically. And then the forest seemed to bend around a shape.

The **mother**.

She was the **largest black bear I've ever seen**, and I've seen hundreds. Easily 300 pounds. She was hopping forward on all fours, snorting, stamping, rising up onto her hind legs. Her mass was overwhelming. Her presence immediate.

It was only through luck and calm that we backed away without an incident. But the moment taught me something valuable:
If you saw her upright, at dusk, from a distance—or just for one second—you might believe you saw something else entirely.

And here's the kicker: that wasn't even close to Ontario's largest black bear.

The **biggest black bear ever recorded in the province weighed over 1,000 pounds**. Now imagine that kind of animal, upright, in fog, or behind half a curtain of foliage. The mind makes fast choices. Sometimes it fills in what it doesn't want to see. Sometimes it substitutes mystery for memory.

That's where some of the confusion comes from.

But not all of it.

Because if this phenomenon were just misidentified bears, **it would've been solved decades ago**.

Part 2: The Hoax Theory — Why Most Hoaxes Collapse Under Their Own Weight

If it's not a bear, then surely it's a prank. A hoax. Someone in a suit, stirring up stories for attention, clicks, or folklore cred.

It's a popular fallback—especially in online spaces. But when you start looking at the details, **the hoax explanation doesn't hold up nearly as often as people think**.

First, the logistics. To pull off a convincing hoax in the wild, you need more than a costume. You need to **get into the backcountry**, sometimes miles from the nearest road. You need to move like a primate, not a man. You need to leave prints that match anatomical features found in **non-human apes**—flexible mid-foot joints, pressure ridges, toe splay. You need to do this on **camera**, or in **mud**, or in **snow**, without ever breaking character. And you need to do it **for no reward**, because most of these sightings never make the news, go viral, or get monetized.

There's no audience in the middle of the woods. Just risk.

I've interviewed people who've made plaster casts of tracks. I've held those casts. Some show **dermal ridges**—the same skin patterns found on fingers and palms. Others show pressure points consistent with something very heavy moving with natural weight distribution. These aren't faked with wood stamps. Not unless you've got a background in biomechanics and primatology.

More than that, most of the **best witnesses aren't looking for attention**. They're former military, hunters, off-duty rangers, bush pilots, loggers. People who know how to move in wilderness and know the difference between **a bear, a man, and something that doesn't belong.** These people rarely report what they see. When they do, it's reluctantly. Quietly. Often with hesitation and anonymity.

And then there's the duration.

This isn't a new phenomenon. Sasquatch sightings date back **well before the internet**, before plastic costumes, before mass media. Indigenous oral histories are full of consistent descriptions —**long arms, heavy walk, strong smell, piercing eyes**—from centuries before "Bigfoot" was a household name.

Hoaxes may happen. Some probably have. But real hoaxes don't last decades. They don't crop up independently across different provinces, cultures, languages. They don't match anatomical consistency across sightings that span thousands of kilometers.

Hoaxes have fingerprints. They carry motives, cracks, confessions.

But when you look at the best cases, **what you see instead is silence**.

A man who doesn't want to talk about it. A woman who breaks down in tears when she tries. A trapper who deletes the footage. A hunter who walks out of the woods and never returns to that spot again.

Those aren't hoaxes.
Those are people who saw something they didn't expect—and still don't fully understand.

Chapter 25: The Smell, The Eyes, The Silence — Traits Described Over and Over

Part 1: The Details That Shouldn't Match — But Do

Across regions, backgrounds, and decades, the descriptions keep overlapping. Not in vague ways, but in **specific sensory detail**—details not easily guessed or imagined. If this were simply mass hysteria or pop culture pollution, you'd expect chaotic variation. But that's not what happens. Instead, you hear the same things.

And you hear them **from people who've never met.**

The Smell

This is one of the most consistent elements reported. People don't just describe an "animal smell." They say things like:

- "Rotten meat and garbage left in the sun."

- "Like a wet skunk dragged through a fire."

- "Burnt hair and sewage."

- "Sickly sweet, like roadkill and mold."

Sometimes the odor comes first—long before the sighting. It arrives **on the wind**, riding ahead like a warning. Sometimes it comes afterward, lingering in a clearing or trail as though something massive just passed through.

One Ontario hunter said he turned around and went back to his truck after twenty years of hunting the same ridge.

"I smelled it before I heard anything. It didn't smell like death. It smelled like something that **chooses to smell that way**. Like it knows you'll leave."

Now it's worth saying—**bears can carry a similar foul smell**. Especially in summer, if they've been eating rot, rolling in carcasses, or if they're wounded or diseased. A big boar in heat or a female guarding a kill can smell absolutely terrible. I've encountered bears that stunk like dead fish, swamp gas, or even worse.

But there's a **difference in character**. People familiar with bears say this other odor is **thicker**, more acrid, as if it doesn't just cling to the air—it pushes into it. One trapper described it as "the smell of something that **wants to be smelled**."

Some skeptics say it could be a large animal with a glandular issue or scenting behavior. But bears and moose don't carry **burnt rubber and ammonia** in their scent profile. And people don't hallucinate smell. It's too visceral, too deeply encoded into survival instinct.

None of these things prove anything on their own. Smells can have explanations. Odors drift. Weather distorts scent trails.

But when strong, distinct odors appear **in remote areas, with no visible animal source**, and are followed by the sense that something massive just passed through?

Then it stops being anecdote.
It starts becoming **pattern**.

Part 2: Sudden Movement, No Sound — How the Impossible Becomes Normal

One of the strangest—and most consistent—details in encounter reports is how these creatures move. People say things like:

"It covered ground way too fast."
"It was just… gone."
"I didn't hear it leave. I should have, but I didn't."

In many cases, witnesses describe a large upright figure, often at the edge of visibility—across a cut line, past a river, or watching from the far end of a field. Then they blink, shift their footing, look back—and it's gone.

But there's no sound.
No branches breaking.

No crashing.
No running footfalls.

Nothing.

You'd expect something that size—8 feet tall, 700+ pounds—to move like a moose in full flight. But that's not what gets reported. What people describe instead is **unnaturally quiet**, fast, efficient motion. Some say it almost "floated." Others say it glided over rough terrain without disturbing anything.

And that's the unsettling part: people don't just describe something big. They describe something **that doesn't move the way big things are supposed to move.**

One man near Temagami was gathering firewood when he saw a dark figure step out from the trees on the far side of a frozen beaver pond. He froze, heart pounding. It stared at him for what felt like a full minute. Then it turned—quick, smooth—and **moved through shoulder-high brush without a sound**. He waited to hear something—snapping, rustling, anything. But there was only stillness.

"It was like watching a shadow decide it didn't want to be seen anymore," he said.

There's a psychological impact to that kind of movement. Our brains are trained to expect resistance—sound, chaos, weight. When that's missing, the result feels **wrong**. It creates cognitive dissonance. People freeze. They doubt themselves. Some don't even speak of what they saw until years later.

Even trained observers—military, law enforcement, hunters—say the same thing. "It shouldn't have been able to move like that." The stride length, the agility, the silence—it doesn't add up.

Could it be a mistake? An illusion of distance or light? Possibly. But when dozens of people, across different regions and backgrounds, report **the same impossible motion**, over and over again?

That stops being a misperception.
It starts sounding like something we haven't classified yet.

Chapter 26: Vocalizations — When the Forest Screams Back

Part 1: Sounds That Don't Belong to Anything Else

You don't have to see one to believe something's out there.

Sometimes, all it takes is standing in the middle of a dark forest and hearing something that doesn't match anything you've ever heard before. Something primal. Something big. Something that doesn't care if you're listening.

Because it's not trying to scare you.
It's trying to **say something**.

Across North America, vocalizations attributed to Bigfoot fall into a few major categories—each of them terrifying in their own way.

The Whoop and Howl

This is the most commonly reported type. A long, rising whoop or bellow that climbs in pitch and volume before fading off or cutting out abruptly. Often it starts in the distance—so far off it almost seems like wind—until it builds into a kind of *siren wail*, echoing across hills, lakes, and valleys. It doesn't sound like a wolf. It doesn't sound like a man. And it's much too powerful to come from a bird.

What's most disturbing is when people hear it **answered**. One whoop from one ridge. A second from another. Then silence. That's not random. That's **call and response**.

I've heard recordings sent to me by loggers, campers, and cottagers who swear there was nothing human nearby. When played back on spectrograms, the sound registers below and above normal human vocal range—**two tones at once**. That's not something the average person can do. That's not something many animals can either.

The Scream

The scream is different. This one stops people cold.

It's often described as part woman, part engine—a scream so high-pitched and violent it seems to tear through the trees. Witnesses often say it triggered **goosebumps, nausea, or an instinct to run**.

A woman I interviewed in Chapleau described it this way:

"It was like something was being murdered. But it just… kept going. Too loud, too long, too big. It was like it wanted everything else to shut up."

Even seasoned hunters have admitted to leaving tree stands early because of this sound. One said he felt it "in his teeth." Another said the woods "never felt right again" after hearing it.

The Chatter

This is more rare. But when it happens, it sticks with people for life.

Witnesses describe fast, garbled speech patterns—something that sounds like **language**, but not any language they've ever heard. High and low tones mixed together. Pauses. Stretches of silence. Then more bursts of rapid syllables.

In 1974, the famous "Sierra Sounds" were captured in the California backcountry—recordings that many researchers still reference today. But similar audio has been recorded in Ontario, Michigan, Washington, and British Columbia.

Some researchers suggest it could be a form of **territorial expression or social communication**, like primate vocalizations in the wild. Others call it "samurai chatter." Whatever it is, it doesn't sound like an animal.

It sounds like something trying to **talk**.

The Best I've Heard

One of the best examples I've ever come across came from a **family near Dryden**. They were out in the bush, filming with a basic camcorder—just casual footage during an evening hike. But what they captured on the audio was something else entirely.

Haunting vocalizations—deep, rolling, mournful—punctuated by distant knocking and a drawn-out wail that echoed between ridgelines. It's raw, chilling, unforgettable. I still believe it's the **clearest example of Sasquatch vocalization ever recorded in Ontario**.

I've listened to that recording over a dozen times.

And each time, it makes the hair stand up on my arms.

Part 2: Who's Listening — The Role of Audio in Modern Bigfoot Research

For years, visual evidence has dominated Bigfoot investigations—photos, films, footprint casts. But in the last decade, there's been a shift. **Audio has become one of the most compelling— and overlooked—tools in the field.**

It's not hard to understand why.

Photographs are fleeting. Tracks can be wiped away by wind or rain. But a sound? A sound recorded in the middle of nowhere, at 3:00 a.m., when no humans are nearby, can't be easily dismissed. Especially when it's powerful, structured, and **nothing else in the area is known to make it.**

The Equipment

Audio recorders have come a long way. What once required reel-to-reel setups or expensive field rigs can now be done with a $200 handheld device or a mobile app synced to a parabolic mic. Some researchers have gone even further, creating custom **long-duration recorders** that can sit silently for weeks, triggered only by noise spikes or unusual frequencies.

I've personally deployed recorders in Northern Ontario. You set it. You leave. You come back weeks later and listen to **hundreds of hours of silence**, hoping to find **five seconds that change everything**.

Modern software lets you analyze those sounds spectrographically—seeing the pitch, shape, and power of the call. You can overlay known species for comparison. Elk, owl, wolf, coyote, loon, human. The ones that stand out? You'll know. They **don't match anything**.

Artificial Intelligence and Audio Patterning

In 2025, things have moved even faster. With access to AI-trained models and large datasets, researchers can now run recorded calls through **machine learning pattern detection** systems.

This means that we can cluster unknown vocalizations—calls that share traits like frequency range, harmonics, and structure.

And what's emerging from that analysis?

Clusters of vocalizations from **different parts of North America**—similar in tone, range, rhythm. Calls from Ontario, Washington, and Appalachia showing **near-identical waveform patterns**, despite being recorded years and thousands of kilometers apart.

That shouldn't happen if these are all random, unrelated animal noises. It suggests a consistent source. A single type of organism. One that can express **territorial warnings, communication, distress, or perhaps even social coordination**.

The Listeners

Then there's the human side. People who hear these sounds don't forget them. And once they've heard them, they begin **hearing them again**, in new places. It's as if awareness unlocks sensitivity. They pick up on tones they once ignored. Patterns they thought were birds, wind, or machinery suddenly **don't sound so right** anymore.

Some researchers now go into the field with **no cameras at all**. Just recorders. Some wear directional microphones like vests. Others hang audio rigs in a radial pattern and sit in the middle. Because once you've heard that sound—whatever it is—you stop thinking of Bigfoot as a myth.

You start listening like it's real,

Chapter 27: The Places They Return — Hotspots That Don't Go Cold

Part 1: Ontario's Recurring Zones

Some places go quiet. A report surfaces, then nothing. No follow-up, no tracks, no sounds. Just a moment in time.

But other places—**they don't stay quiet**. They keep surfacing. Year after year. Decade after decade. The same regions, the same descriptions, the same uneasy feeling. In some cases, even the same trails. It's like something lives there. Or returns to it regularly.

These aren't vague rumors. They're **documented patterns**. And Northern Ontario has more than a few.

The Temagami Triangle

This stretch of rugged bushland, lying roughly between North Bay, Temagami, and the edge of the Lady Evelyn-Smoothwater Park, is remote, wild, and full of waterways. And for over 40 years, it has been the source of recurring sightings, audio events, and strange track discoveries.

Local guides talk about **no-go zones**—places where even seasoned trippers won't camp anymore. You hear about trails that "feel wrong," places where wildlife goes silent, and shadows seem to move uphill. One logging crew reported a series of enormous prints along a pipeline cut, spaced like a tightrope walker moving through mud.

The area is too remote to sustain a hoax. Too far from roads to be a prank.

And the reports keep coming.

Manitoulin and the North Shore

Between Espanola and the north end of Manitoulin Island, several small rivers cut through thick forest and swamp—perfect travel corridors for animals. And maybe something else.

Fishermen, campers, and local hunters have all reported **distant vocalizations**, rock throws, and tree shakes. Some say they've seen dark figures cross roads at impossible speeds. Others talk about finding stick structures where there shouldn't be any—woven limbs bent in patterns not caused by wind or snow.

The smelt run in spring draws people to the water's edge—and it may draw other things too. I investigated a **case just off Highway 6**, where two smelt fishermen witnessed a **boulder the size of a basketball thrown across a river**. No warning. No sound. Just a sudden splash and the feeling of being watched. It was dark. They were alone. And they left immediately.

The Hearst-Hornepayne Corridor

Dense spruce, narrow logging roads, and almost no population. This area doesn't get many hikers or cottagers. But it gets reports.

Trappers have mentioned **heavy movement in the trees**, long beyond the size or sound of moose. A bush pilot described seeing a **lone figure walking upright along a fire break**, so remote there were no signs of vehicle entry for miles. The pilot circled twice. By the third pass, it was gone.

A forestry crew once found a section of **bark peeled off twelve trees**, about seven feet up— freshly done, in a straight line, with no claw marks or machinery near. They didn't report it officially. But one of them took photos. He didn't know what to make of it. Just said it felt like someone was showing off.

West of Thunder Bay

Closer to the Minnesota border, where deep lake systems connect to vast stretches of Crown land, a different kind of pattern emerges: **aggression**.

Reports from this region sometimes include rock throws at tents, wood knocking close to camp, and low growls from just beyond the firelight. One case involved a pair of night hikers who had their trail blocked by a **deadfall that hadn't been there twenty minutes earlier**—deliberately laid across the path, still weeping sap.

They left. Fast.

Part 2: The U.S. Zones — Consistent Hotspots South of the Border

Canada isn't alone in these patterns. Across the United States, certain regions have produced steady reports going back decades—**long before television shows or online maps ever made Bigfoot sightings mainstream.** These are not one-off encounters. These are **geographic echoes**—the same behaviors and descriptions, returning generation after generation.

And when you look closely, you start to see how the terrain itself invites them.

The Olympic Peninsula, Washington

Dense rainforest. Constant rainfall. Steep ridges and winding rivers. This stretch of coastal wilderness has long been a centerpiece for Sasquatch research—and for good reason.

Reports from the Olympic Peninsula often include:

- Extended vocalizations from ridgelines.

- Upright figures crossing roads late at night.

- Tree knocks echoing through fog.

- Massive tracks found in muddy streambeds.

Jeffrey Meldrum, Ph.D., has examined multiple track casts from this area and notes strong **midtarsal break morphology**, suggesting an entirely different foot structure than humans or bears. Some of the best high-detail casts ever collected came from this region.

The Quinault and Hoh tribes also have generations of oral history that mention **"forest people"**—tall, silent watchers who live near water and avoid human contact.

The Green Swamp, Florida

The idea that Bigfoot would live in Florida might seem strange—until you understand the **scale and isolation** of the Green Swamp.

This 500,000-acre region west of Orlando is a labyrinth of thick palmetto, floodplain, and heat-choked pine scrub. It's inhospitable to humans, but **perfect cover for wildlife**. And people have been reporting **skunk-ape encounters** here for decades.

Descriptions often include:

- Strong, putrid odor (hence the name "skunk ape").

- Broad, upright figure glimpsed at dusk.

- Large tracks in wet sand or near swamp edges.

Many of the reports come from **rural locals**, not tourists. They're often unreported—shared only through private forums, podcasts, or passed between hunters.

The Ouachita Mountains, Arkansas & Oklahoma

Heavily wooded, steep terrain. Thin population density. And a long, long history of **booger** legends—local slang for hairy wild men that predates the term "Bigfoot."

A 2021 case involved a series of **whoops and screams** recorded over five nights near a deer lease. Audio analysts described the recordings as "inconsistent with known species and too low-frequency for a human source." A separate sighting from 2019 involved two bowhunters who watched a **tall, wide figure crouched near a creek**, then rise and walk into the trees without a sound.

These are places where **the same trails get used year after year**—by people, animals, and maybe something else.

The Allegheny National Forest, Pennsylvania

People don't always associate the East with Bigfoot activity, but **western Pennsylvania is loaded with cases**—especially in the area surrounding the Allegheny River and the national forest lands.

In these deep ravines and old-growth valleys, researchers have recorded:

- Wood knocking patterns.

- Tree breaks at shoulder height.

- Silent shadows watching from distance ridges.

And the locals know. They don't always talk about it. But they know.

From Ontario to Oklahoma, from coast to coast, **something is returning to the same places, over and over again**. The evidence doesn't sprawl randomly—it clusters. It settles. It comes back.

Chapter 28: The Ones Who Vanish First

Part 1: Animal Behavior During Sightings

Some people think the forest goes quiet before a storm. I know better. In the places where the strange happens—where the stories cluster and the hair on your neck stands up—the quiet comes first. Then the encounter.

But it's not always a person who reacts first.

Dogs freeze. Horses buck. Moose bolt. Chickadees vanish like a switch has been thrown. You could call it instinct. You could call it something else. I've lived it. I've measured it. And in the places where Bigfoot has been reported—especially in Northern Ontario—it's often the animals who feel it first.

I've seen it in my own dog—my German Shepherd, Chase. He isn't easily rattled. But I've watched him stop dead in the bush for no reason I could see. No visual. No sound. Just a locked-in posture, ears angled, body stiff. Not aggression. Not fear. Something else. Something closer to uncertainty. As if the thing he sensed was calculating.

That's happened more than once. Two occasions stand out—both on scouting walks, both in areas with recent vocalization reports or track impressions. The first was in late May north of Timmins. The other was deep in Algoma District, mid-October. In both cases, Chase halted without a trigger. I scanned, I listened, I moved. He didn't. Not until we turned back.

Other handlers have told me similar things. A hunting guide I know in the Haliburton Highlands said one of his dogs wouldn't pass a certain trail bend—something it had done dozens of times before. No clear reason. No scent I could detect. But the dog wanted nothing to do with it. A trapper near Hearst told me all three of his huskies turned and walked back to the truck before he'd even noticed the forest had gone quiet.

And that silence is worth paying attention to.

They call it "the dead zone." A patch of forest that should be alive with bird calls and squirrel chatter, but isn't. Not because of weather. Not because of time of day. But suddenly, sharply, and all at once. Like walking into a room with the air turned off.

I've stepped into it more than once—places where fifty feet in, the world just stops making sound. No wind. No insects. No breeze in the pines. Just an unnatural stillness that grips the back of your neck. Sometimes there's pressure in your ears, like altitude shift. Sometimes you feel it in your gut before you even register what's wrong.

And then there are the horses.

In the Rainy River District, a woman who'd raised horses for more than twenty years described one such moment. It was dusk. Her animals—normally calm—started to snort and back toward the barn. Not spooked, but moved by something. She didn't hear anything at first. Just felt the energy shift. Then came a single, heavy knock from the woods behind the paddock. Her border collie flattened itself to the ground. Not barking. Not curious. Just prone. And it stayed that way until the horses were back inside.

She told me she hadn't shared the story for years. "It sounded crazy," she said. "Until I heard someone else describe the knock."

It's not proof. But it's pattern.

When people talk about sightings, they describe what they saw. But the animals—if you pay attention—show you what's approaching before it ever appears. They don't always react the same way. Some growl. Some cower. Some disappear. But when it happens in conjunction with forest silence, that's when I start to pay attention.

Infrasound? Scent? A sixth sense we don't have? I'm not claiming answers. Just observations, over time, in real bush settings. And I can say this: if your dog stops, your horse turns, and the forest goes quiet—don't call it coincidence.

Something is watching. And your animals know it before you do.

Part 2: Predator Logic — When Bears, Cats, and Wolves Don't Match the Signs

You spend enough time in the woods, and you start to catalog predators. Not just by tracks, but by atmosphere. Bears move differently than wolves. Wolves don't sound like cats. The bush reacts to each one in a specific way. You can feel it—especially when you've seen it repeat year after year.

When there's a bear nearby, you know it. Not always because you see the animal, but because of the collateral signs. Cracked branches, overturned logs, pungent scent—often like ammonia or rotting greens—especially in the hotter months. You get motion. You get sound. You get response from the smaller wildlife. Crows change tone. Jays scream. Insects move.

But what happens in some of these Bigfoot-associated incidents doesn't line up with known predator logic.

A bear doesn't cause birds to vanish in an instant. Even the largest black bears rarely draw prolonged attention from adult moose or herds of deer unless it's a sow with cubs. Predators are part of the ecosystem—everyone knows the rules, even the prey. They assess the risk and move accordingly.

What I've experienced in the so-called "quiet zones" feels different. It's not the calculated caution of prey facing a known threat. It's a kind of unnatural stillness. Not avoidance. Not retreat. But absence.

I've been in bear country when it's alive. I've also been in places where people claimed to have recent Bigfoot activity. The difference is stark. With a bear, the forest leans away from the animal. With this other thing, it disappears entirely.

I've tracked wolves in fresh snow. You can see the spacing of the stride, the pattern of drag or dig when they move quickly. You can often hear their approach well before you ever see them. With cougars, the signs are finer—scrapes, tree markings, territory passes. But they don't leave entire patches of bush dead silent. And they certainly don't cause multiple animals to refuse to proceed past a certain invisible threshold.

There was a case in early winter just west of Wawa. Two hunters reported a moose bolt across a logging road at full speed. Behind it—nothing. No sound. No scent trail. But their pointer dog—seasoned, reliable—whined and refused to cross the trail. It tucked low, tail down, not frightened but deeply unsettled. The hunters checked the treeline, listened, and waited. Nothing came. No second animal. No crack of underbrush. Just nothing. It was like whatever had spooked the moose had stepped into a different layer of air.

These are the stories that don't make headlines. They don't end with video or howls or visible figures. But they show a pattern. And when those patterns contradict everything we know about Ontario's known apex predators, I pay attention.

I've had other researchers tell me something similar. That whatever's happening in these areas, it isn't behaving like a predator—it's behaving like something the animals can't categorize. Something they don't run from right away. Something they watch. Judge. Then quietly choose to avoid.

And in the forest, avoidance without threat is a red flag. Because it means the animals aren't reacting out of fear. They're reacting out of unfamiliarity.

That's not how bear encounters feel. That's not how wolves move through dense bush. And it's definitely not how known wildlife makes decisions.

Part 3: The Second Witness — When Animals Confirm the Unseen

Some of the best witnesses don't speak. They don't rationalize what they saw. They don't downplay what they heard. They react. And they remember.

I've come to think of them as the second witnesses—dogs, horses, even livestock—that behave differently in the presence of something out of place. And not just in a flinch-and-go way. I mean sustained, deliberate behavior that mirrors what a human witness feels but doesn't always trust.

I've interviewed people who barely remember what they saw—but remember exactly what their dog did. People who weren't scared until the animal turned and ran. Not bolted. Ran. Tail down, ears flat, no sound.

A man outside Temagami told me about a summer night camping with his teenage sons. They had two dogs with them—labs, bush-smart, used to camping. At around 1 a.m., the dogs started pacing the tree line. Not barking. Just moving back and forth, heads low, growling so faintly he didn't hear it until he turned off the radio. When he finally stepped outside the tent with his flashlight, the dogs were both staring at the same point in the trees, muscles locked. He couldn't see anything. But when he stepped toward the bushline, both dogs backed up in sync, as if to block him from going further. He told me that rattled him more than anything. "They weren't scared," he said. "They were trying to stop me."

Another case from Rainy Lake: a couple walking their property in late fall came across their dog —an older mutt with no history of spooking—sitting at the edge of the tree line, body shaking. Not from cold. From something else. The husband called the dog three times. It didn't move.

When it finally came, it wouldn't go past a certain point. No prints. No scent. But the woods felt wrong. The man said he couldn't hear a thing. "It was like someone turned off the sound around us," he said. "And my dog knew it before I did."

There are variations of this story all over the north. Dogs growling at silent, empty forest. Livestock refusing to graze near a certain section of field. Horses backing away from treelines without cause. The details change—but the shape of the story doesn't.

That's the thing about field research. You can measure scat. You can cast tracks. You can run audio gear until the batteries freeze. But it's the small details, the reactions of other living beings, that sometimes tell you the most. Because animals don't get confused by their own fear. They don't worry about what they're "supposed" to have seen. They don't interpret. They respond.

And when those responses come in clusters—when multiple dogs refuse to walk a trail one week after a local sighting, or when a horse has to be dragged out of a barn for days after a strange noise—it starts to add up.

I don't use animal behavior as proof. I use it as compass. If your dog stops walking where the bush goes silent, I listen. If a seasoned hunting hound tucks tail in a zone known for strange reports, I take notes. If someone tells me their horse won't face a clearing anymore after one particular night—I mark the map.

Animals don't chase ghosts. They respond to what's real.

And if they're reacting to something we can't see?

Then it's not just our mystery anymore. It's theirs too.

Chapter 29: Where They Move — Corridors, Seasons, and the Northern Flow

Part 1: Old Trails, New Questions

In Ontario, the land speaks in lines. Old trapper paths. Forgotten logging roads. Game trails that wind like veins through granite and muskeg. Some of them haven't seen boots in fifty years—but something's still using them.

I started to notice it after my third year of field work. Reports weren't just random dots on a map. They followed shape. They hugged the edges of lakes, mirrored the old Algonquin trade routes, and ran parallel to ancient glacial pathways. It was more than coincidence.

Northern Ontario is massive, but not featureless. The Shield splits it with rocky ridges and low valleys. And what I began to see, again and again, was that sightings and evidence—the best ones—came from transition zones. Places where one type of terrain met another. Forest edge to bog. Swamp to rock. River to hill.

A Cree elder from the James Bay coast once told me something I've never forgotten. "The animals move where the land gives them silence." I think about that every time I map a trackway or follow up on a new report.

Take the cluster west of Thunder Bay. Multiple reports, spread across two decades, nearly all aligned along the same north-south curve. Not towns. Not roads. Landforms. The path follows a natural lowland trench—one that runs uninterrupted for nearly 90 kilometers. It's invisible to most people. But on foot? You'd feel it. And if you were something trying not to be seen, it would be the perfect trail.

There are places like that all across Ontario. Unbroken lines through unbroken land. Not visible from highways. Not marked on the usual maps. But alive with travel, with movement. I've had trappers tell me they've seen the same deep tracks year after year along the same creekbeds, always when the snow just begins to fall.

Not wandering. Following.

There's a theory among some researchers that what we call Bigfoot might be migratory—not like birds, but in a slower, seasonal sense. Summer highlands, winter retreats, predictable loops repeated across generations. The farther north you go, the more that makes sense. Winters here can break a moose. A creature without strategy won't last.

And then there's the Lake Superior loop.

Some of the most convincing evidence from both Canada and the U.S. clusters around the edges of Lake Superior, forming a slow arc from Duluth to Wawa and back again along the northern shore. Trackways near Terrace Bay. Audio from Nipigon. A Class A sighting off a canoe route east of Marathon.

They're not close together by car. But by land? By foot? They form a loop. A route. One that keeps to ridgelines, water access, and old growth cover.

It's the same across the continent. In the Pacific Northwest, track reports hug the same coastal ridges year after year. In the Appalachians, sightings cling to the spine of the range. And in Ontario, you start to see the shape emerge only after enough time, enough reports, and enough walking.

Because the shape is movement. And movement is the only way something that big stays hidden.

Part 2: The Winter Retreat Theory

You don't last long in Northern Ontario winters without a plan. The cold isn't just uncomfortable —it's dangerous. It locks up food sources, silences running water, and drives nearly every large mammal into a survival routine. Moose yard up. Bears den. Even wolves conserve movement. But the reports don't stop.

In fact, some of the clearest evidence—the most baffling trackways and unsettling audio—comes from deep winter. Subzero temperatures. Waist-deep snow. The kind of conditions where no hoaxer's going barefoot for miles, and no known predator is wandering open country without cover.

That raises the question: If Bigfoot is real, what does it do in winter?

The popular image is a mountain-dwelling creature, built for elevation and cooler climates. But Ontario isn't like the Rockies. Our highlands are dense, uneven, and lack the vertical escape routes of western ranges. Survival here, in the coldest months, means knowing the land on a granular level. Where the thermal pockets are. Which valleys hold spruce cover. Where the marsh doesn't freeze solid. Where the deer pass through.

And I think that's exactly what it does.

One January, near the Chapleau Crown Game Preserve, a snowmobiler found a trackway in fresh powder that led out of a heavily wooded gully, crossed a hydro cut, then disappeared into a tangle of cedar lowland. The prints were spaced over four feet apart. Each was nearly sixteen inches long. The area had no recent human traffic. No cabins. No hiking routes. Just snow and bush and power lines.

I walked it days later. The snow was still deep and clean. Moose tracks were everywhere. So were the signs of other large predators—fresh wolf tracks, scat, and bedding spots. But none of those trails intersected the trackway. Not once. As if everything else gave it a wide berth.

A local outfitter I trust told me he thinks these things follow the same idea as wolves: find a food-rich zone in winter and stay within it. But where wolves leave vocalizations, territorial markings, and frequent tracks, these don't. Whatever they are, they're moving with more caution—or more understanding of how to go unnoticed.

One theory I've considered—based on multiple sightings from January to early March—is that they shift into known game corridors. Places where deer, elk, and moose bunch up in deep snow, moving slowly and predictably. These areas aren't always visible on maps, but they're known to locals—especially trappers and First Nations communities.

I've even heard stories of makeshift dens. Caves reused each season. Rock overhangs packed with spruce boughs. One snowshoer north of Elliot Lake came across a deep depression in the snow under a cedar thicket—no visible tracks in, but the snow was melted in a circular bed nearly eight feet across. Steam still rising from the center. He left without checking further.

Winter doesn't stop this creature. It reveals it.

The snow becomes a canvas. The cold narrows the options. And that's when patterns show up—when movement becomes easier to detect. It's also when the thing, whatever it is, seems to be most rooted.

Not passing through.

Staying.

Part 3: Mapping Memory — Why They Return to the Same Places

Some sightings aren't new. They're echoes.

A story comes out—something recent, local, detailed—and if you listen long enough, someone else steps forward. Not with something similar. With something identical. Same spot. Same time of year. Same behavior. Only it happened thirty years earlier.

I used to think that was just coincidence. Or memory bias. But after tracking dozens of these layered reports, I don't think so anymore.

I think these creatures—if they're real—remember.

And not just in an instinctual way. In a mapped, seasonal, deliberate way. Like migratory birds returning to the same nesting ledge. Like bears revisiting the same berry stands year after year.

In Northern Ontario, the bush doesn't change fast. Logging comes and goes. Lakes freeze and thaw. But the old corridors remain. And I've come to believe some of these creatures use those corridors with purpose. Not by accident. Not by wandering. But by memory.

One man near Dubreuilville reported tracks in the same section of fire-cut trail two winters in a row. No human activity. No hunting in that zone. Just a quiet, curving piece of brush path with flat snow on either side. Same direction. Same spacing. Same print size. He didn't tell anyone the first year. But he marked the date. And when it happened again, nearly to the day, he came forward.

Another case came from a woman near Vermilion Bay. Her father had seen something in the late '70s—a tall figure moving upright through a marsh behind their property. It stopped, looked toward the house, then vanished into tamarack. Forty years later, her daughter—same yard, same time of year—heard a low, steady knocking one night and saw something massive pass between the trees at the edge of the floodlight. She never knew about her grandfather's sighting. Not until she described the exact spot where it happened.

That's not random. That's route memory.

Animals use memory differently than we do. They don't have nostalgia. They don't mark anniversaries. But they do remember where the food was. Where the shelter was. Where they weren't seen.

A creature smart enough to avoid people for decades—smart enough to hide tracks, control sound, pick the wind direction—would almost certainly remember safe zones. And if it passed that knowledge down?

Then what we're seeing isn't just evidence of a population. It's evidence of *patterned movement* —a behavior passed between generations.

And that changes everything.

It means some sightings aren't accidents.

They're returns.

Chapter 30: Technology in 2025 — What's New, What Still Fails

Part 1: Better Gear, Same Silence

We have better tools than we've ever had. Trail cameras with AI motion sorting. Drones with thermal imaging. Microphones that can pick up a whisper from a kilometer away. The kind of tech that should make mysteries like this impossible.

And yet—nothing.

Nothing definitive. No irrefutable footage. No thermal walk-through that stuns the scientific community. No long-range audio that ends the debate. Just a better class of silence.

I've used all of it. Trail cams with passive infrared sensors. Units designed to detect not just movement, but body heat. I've mounted them facing bait stations, game trails, mineral licks—anywhere large mammals might pass. I've used scent blockers. Camouflaged the cases. Buried them halfway into rotting stumps. I've even timed deployments to within hours of reported sightings.

I've captured moose. Bear. Wolves. Illegal loggers. Poachers. Lost hikers. I once caught a white moose—rare as they come—ambling through a pine stand just after dusk.

But never a Bigfoot.

It's not just me. I've talked to researchers across North America with similar stories. People who've deployed thousands of dollars in gear. Hours of setup. Dozens of units running simultaneously, triangulated across known hotspot zones. And still—nothing. Or worse: corrupted files. Blanks. Batteries dead the morning after a vocalization was heard. Cameras working flawlessly until the one night something actually happened.

The same goes for drones.

We now have quadcopters that can scan entire forest grids in minutes, flying low and quiet, equipped with heat-mapping tech originally designed for search and rescue. But flying them in dense boreal forest isn't like using them over farmland or townships. Tree canopy messes with visibility. Cold affects battery life. And while thermal contrast works well on open ground, it's easily lost in spruce thickets or cedar marsh.

We've flown them. I've flown them. Over trails where fresh prints were found, in valleys where calls were recorded, above ridge lines where something moved too fast and too upright to be a bear. The footage looks promising—until you actually need it to prove something.

Same with audio. We have software now that can isolate sound signatures, map their decibel curve, and compare them to known species. I've personally run vocals through spectrographs that show sustained pitch, diaphragm volume, and midrange formants that don't match any local animal. But it's still not "proof." Just another anomaly.

So what does that mean?

It means the problem isn't gear. It's behavior.

We're trying to record something that doesn't want to be recorded. Something that doesn't blunder through a camera trap like a curious bear. Something that moves with timing, intention— and maybe even a form of low-level surveillance awareness. It sounds wild. I know it. But the patterns are too consistent.

Maybe it's not just random stealth. Maybe it's a form of learned avoidance.

Whatever the case, we've crossed into an age where lack of evidence can't be blamed on bad equipment. We're seeing the edge of a truth we still don't understand.

Better tools. Smarter setups. Same result.

And if that doesn't make you rethink the mystery… nothing will.

Part 2: AI, Glitches, and What Cameras Miss

A few years ago, a group in Minnesota ran a deep-learning filter across their entire archive of trail cam footage—hundreds of hours, thousands of frames. The system was trained to detect anomalies. Not just animals, but shapes, gait patterns, speed deviations. It was supposed to flag anything that didn't match known species behavior.

It flagged 137 clips.

Of those, 131 were immediately dismissed—wind shadows, birds too close to lens, glare. Four were just system glitches. One was a bear moving upright on hind legs. And one?

Unresolved.

A blur, dark, two frames only, between pine trunks on a cold morning in February. The size was wrong for a bear. The proportions didn't match human. The system logged it as "non-classifiable biped."

That was it.

Not evidence. Not proof. Just a hiccup in a system designed to spot the unexplainable—and then bury it in a folder no one would look at twice.

That's the other problem with 2025 tech. It's not just good—it's blind in new ways. AI image filters are built to discard the unexpected. Their whole job is to recognize *known* patterns. Anything outside that gets tossed, flagged, or mislabeled.

One researcher I know from British Columbia caught a strange figure on a trail cam—tall, in motion, just at the edge of frame. The image recognition software tagged it as "tree movement." If he hadn't been reviewing every file manually, it would've been deleted automatically.

That's not science fiction. That's current workflow.

Even civilian camera phones do it. Night filters smooth out edges. HDR algorithms flatten shadows. Motion stabilization deletes the blur of something moving fast. It makes forest photography easier to digest—but less honest. The very systems designed to help us see more are quietly erasing the very evidence we're after.

It's one reason I still insist on reviewing raw footage. Uncompressed files. Manual inspection. Multiple angles if possible. Because when the tech gets too smart, it starts deciding what we're *supposed* to see.

And Bigfoot, if it exists, is the kind of presence that doesn't sit well with "supposed to."

I've seen high-end camera rigs placed at active sites—motion-triggered, weather-sealed, and connected via cellular signal to live-feed databases—fail for no explainable reason. One had perfect battery and function for weeks, then shut off the night after fresh tracks were found. The

SD card was intact. No damage. Just 18 missing hours of data. The only night anything strange happened.

Was it a glitch? Maybe.

But I've learned to take glitches seriously.

The truth is, the more advanced our tools become, the more we start assuming the *tools* are the solution. But they're only as good as what they're looking for. And when what we're looking for doesn't want to be found?

The smartest lens in the world still misses it.

Part 3: The Human Problem

It's easy to blame the mystery on gear. Faulty cameras. Bad angles. Corrupted files. But the deeper I get into this work, the more I realize the failure isn't just technical—it's human.

Because even when the gear works perfectly, we don't.

We second-guess. We dismiss what doesn't fit. We edit out the strange. We chalk it up to weather, light distortion, lens flare. And even when we do capture something strange, we hesitate to share it. Not out of fear of the creature—but fear of ridicule.

I've spoken with landowners who've sat on clear, unexplainable footage for years. Home security cams that caught massive figures crossing snow-covered driveways. Audio recordings with no match in known animal databases. You ask them why they never showed anyone and the answer's almost always the same:

"I didn't want to be laughed at."

That hesitation—that instinct to disbelieve what we see with our own eyes—is the real barrier. Because in 2025, we've created a paradox: the better our tech becomes, the more we expect it to explain everything.

If it doesn't match an algorithm, we throw it out. If a photo looks too clear, we assume it's fake. If a video is blurry, we say it's not good enough. There's no winning standard anymore. Just a moving goalpost of disbelief.

Even among researchers, the human problem lingers. Ego. Competition. The desire to be first, or right, or validated. Instead of building a cooperative network, we have splintered groups, each guarding their data like currency. Valuable sightings vanish into personal archives. Audio files get buried under NDAs. Locations are redacted beyond usefulness.

And then there's confirmation bias.

I've seen people convince themselves that a raccoon is a juvenile Bigfoot. I've watched thermal footage get misinterpreted by well-meaning but inexperienced eyes. I've seen the reverse too— clear, strange movement dismissed because it doesn't match the *idea* of what a Bigfoot should look like.

Tech doesn't fix that. It magnifies it.

The real issue isn't just whether we have the tools. It's whether we're honest enough—humble enough—to interpret what they give us without twisting it to match our expectations.

We've entered an age where the gear is no longer the bottleneck.

We are.

Chapter 31: What We Know Now — 70 Years Later
Part 1: The Pattern in the Static

Seventy years. That's how long the search has been serious—organized, methodical, documented. Not just folktales around a fire, but boots on ground, names on maps, casts taken, audio logged. From the first wave of sightings in the Pacific Northwest to the modern reports rolling in from trail cams and TikTok posts, the Bigfoot question has outlived trends, skeptics, even some of the original researchers.

And yet, the answer remains elusive.

But here's what's changed: we now have patterns.

When you strip away the noise—the hoaxes, the blurry roadside stories, the people chasing clicks—you're left with data. And that data doesn't lie. Across time, location, and culture, a few things have become stubbornly consistent.

1. The Description
Eight to ten feet tall. Hair-covered. Upright. Wide shoulders. Long arms. No neck. Eyes sometimes glowing. Movement smooth, quiet, incredibly fast for its size. From the swamps of Florida to the ridgelines of British Columbia, the core traits haven't shifted. The names change— Sasquatch, Skunk Ape, Windigo—but the description holds.

2. The Behavior
Avoids contact. Watches from cover. Follows from a distance. Approaches camps at night. Taps on windows. Mimics voices. Throws stones. Breaks branches in patterns. Leaves when noticed. Not aggressive—deliberate.

3. The Environment
Thick bush. Water nearby. Low human presence. Deep ravines. Swamp corridors. Cedar stands. Always somewhere a man wouldn't wander alone at night. And where reports cluster, game is usually abundant.

4. The Timing
Peak sightings happen during early spring and fall—times of movement. During winter, fewer reports, but more tracks. Summer brings more human interaction, but also more hoaxes. The real data follows migration logic.

5. The Witnesses

Not all drunks or dreamers. Many are experienced outdoorsmen, trappers, hunters, First Nations elders, police officers, pilots, biologists. People who know the bush. Who know what moose tracks look like. Who know the difference between bear vocalizations and something that sounds like a human scream thrown through a megaphone.

That's what seventy years has given us: not proof, but *weight*. A mountain of stubbornly consistent details, delivered by people who never asked to be believed, only listened to.

And if we're honest—if we truly believe in science as a process, not a dogma—then we have to admit that something is happening out there.

Not a trick of light. Not a shared hallucination. Not hundreds of unrelated liars across decades and continents.

Something.

Part 2: What's Still Missing

We don't have a body.

That's the first and final argument from skeptics. No bones, no specimen, no DNA match that can't be explained away. Seventy years of stories, tracks, and howls—and still no smoking gun.

It's a valid point. Science depends on verification. A new species requires physical evidence. Not blurry footage or emotional interviews. Flesh, blood, and bones.

But here's what people forget: lack of evidence doesn't equal evidence of absence. Especially in the wilds of North America. Especially in the boreal forests and alpine zones where nature covers its tracks better than any man.

We've lost planes in Ontario. Entire aircraft. Never found. We've lost people less than a mile from highways, never recovered. Bears die in the woods and vanish within weeks—scavenged, scattered, hidden beneath new growth. And bears don't avoid open space like these creatures do.

Then there's the DNA question.

We've collected hair. Saliva. Skin oil from tracks. Some tests come back as "unidentified primate," others as "human contamination," and a few as "no viable sample." The problem isn't always the sample—it's the lab. Most facilities aren't equipped to test for a species they don't believe exists. The unknown gets shoved into the "inconclusive" drawer.

Even the best researchers have hit walls. Melba Ketchum's controversial DNA study drew attention, but lacked academic support. Other promising leads disappeared into legal disputes, withheld samples, or credibility battles within the field. It's not just the science that fails—it's the politics around it.

And the photos? The videos?

We've seen hundreds. Thousands. Most are easy to dismiss—costumes, misidentified wildlife, even clever editing. But a few… a very few… are different. The 1967 Patterson-Gimlin film remains the gold standard—not just for clarity, but for the way the creature moves. Its gait. Its proportions. Its musculature. No suit could replicate that—not with 1960s tech. And no human walks like that, with that heel-to-toe motion and fluid hip rotation.

Newer footage exists, some even more compelling, but most never see wide release. Why? Because no one wants their name destroyed in the court of public ridicule. Because even now, to say "I saw it" is to risk your career, your reputation, even your safety.

What we're missing isn't just proof.

We're missing *permission*—for people to speak honestly without being mocked. For scientists to investigate without risking tenure. For footage to be analyzed without being dismissed on sight.

We don't have a body. But maybe the reason is because we're not allowed to see it—because the moment someone does, the rest of the world rushes in to say it didn't happen.

Until that changes, we'll keep circling the truth—closer than we've ever been, yet still just out of reach.

Part 3: The Evolution of Belief

When this all started—back in the '50s and '60s—belief in Bigfoot was simple. You either thought the creature was real, or you didn't. It was a yes-or-no question. One path led to adventure, campfire stories, plaster casts, and newspaper clippings. The other led to eye rolls and laughter.

But belief has changed.

It's more complicated now. More fragmented. More cautious. After decades of ridicule, many serious researchers no longer use the word *Bigfoot* in their reports. They say "anomalous primate" or "upright hominid." Not to be coy—but to avoid the cultural baggage that comes with the name. The cartoons. The T-shirts. The tourist traps.

The mystery has outgrown the meme.

What we know now is that this isn't just about whether the creature exists. It's about what kind of animal could exist, unnoticed, in modern North America. It's about biology, yes—but also psychology. Anthropology. Linguistics. Pattern recognition. Human error. Fear.

And somewhere beneath all that—something real.

Some believe Bigfoot is an undiscovered North American ape. Others think it's a relict hominin —something closer to us than we're comfortable admitting. A few insist it's paranormal or interdimensional, though I don't share that view. I believe it's flesh and blood. A physical creature. Elusive, intelligent, but real.

What matters most is this: after seventy years of sightings, trackways, sounds, and first-hand reports, belief is no longer fringe. It's shifting into something closer to cultural consensus—not mainstream, not settled science, but a growing undercurrent of "Maybe."

And that "maybe" is everything.

Because belief doesn't always come from evidence.

Sometimes it comes from patterns.

From stories whispered across decades that line up too closely to ignore. From hunters who swear by what they saw but will only tell you over a campfire, away from phones and notepads. From biologists who *almost* go public, then don't. From families who find 18-inch tracks in fresh snow and never speak of it again.

These are not believers. They are observers.

And when you have enough observers… belief becomes something else.

It becomes **probability**.

It becomes **momentum**.

It becomes the slow turn of a world waking up—not to fantasy, but to the possibility that we've missed something big.

Chapter 32: Why We Keep Looking

Part 1: The Chase Is the Answer

If you've made it this far, you know the truth.

Not the answer. Not the proof. But the truth behind the question.

Because chasing Bigfoot isn't just about solving a mystery. It's about the chase itself. The discipline it demands. The humility. The discomfort of walking into dark woods with no guarantee, only the hope that you'll come back with something that matters.

Most people don't understand that. They think we're chasing shadows. They think it's about the adrenaline, the fame, the idea of discovering a monster in the trees.

But for those of us who've been out there—who've stood in remote forests at 2 a.m. with our breath freezing midair and our recorders running—it's never been about *finding* Bigfoot.

It's about *listening* for him.

It's about learning how to read a trail the way you read a language. How to tell when silence isn't just quiet, but intentional. It's about standing in the place where something ancient might have passed, and trying to imagine what it left behind.

Why do we keep looking?

Because every now and then, the forest speaks back. Not with words. Not with evidence that would hold up in court. But with a feeling. A shape. A sound. A fleeting presence that resets everything you thought you knew about the world.

You go in thinking you're looking for an animal.

You come out realizing you're looking for *truth*—about what still hides, what we've ignored, what we've chosen to pretend isn't there.

Part 2: The Last Footprint

This search—it's one of the last true adventures left on this continent.

We live in a world where satellites can read a license plate from orbit, where drones can scan valleys in seconds, where GPS can track our every step. And yet—despite all of that—there are places in Northern Ontario, in Washington, in the deep Kentucky hollers and the dark Carolina swamps, where mystery still breathes. Where something big still moves just out of reach.

That's why we keep going back.

Because no matter how many tracks we find, or how many vocalizations we record, there's always the next ridge. The next valley. The next impossible-to-explain shape glimpsed through trees. And because there's still a part of us—not just as researchers, but as human beings—that craves the unknown.

This isn't just science. It's wilderness. It's modern-day exploration. It's real.

The Bigfoot search is one of the last unscripted journeys left—where phones don't always work, where the terrain can still humble you, where you can walk for hours and never see another person. And when you do catch a sound, or find a single print in fresh snow... it's not just exciting.

It's *meaningful*.

It means the world hasn't given up all its secrets. It means there are still things we don't understand. It means that wonder—the thing we were born with as kids but taught to set aside as adults—might still be earned. Not with belief, but with effort.

And yes, maybe the destination is blurry. Maybe the answer will never come. But the *journey*— the actual time in the field, the stories traded around a camp stove, the long drives to trailheads, the nights when the woods go silent and you don't breathe until they start again—that's what matters.

The Bigfoot search forces you to pay attention.

To the dirt. The trees. The sound of your own boots. To the air pressure before a call. To the shift in crow chatter. It turns every walk in the woods into a field lab. Every night into a test of patience and nerve.

Some people say, "If you haven't found it after all this time, maybe it's not there."

But maybe the finding isn't the point.

Maybe the point is that we *still go*.

Part 3: What Comes Next

There's a moment that happens after every expedition. After you've packed up the gear. After the batteries are dead and the thermals are cold and the trail cams are silent. It's quiet. Just you and the bush.

And that's when it hits you.

The question isn't "Did I find it?"
The question is: "Will I go again?"

For most of us, the answer is always yes.

We go again.

Because we know that the next call might come from a farmer who swore he saw something cross his field at dusk. From a trapper who found tracks where there should be none. From a pilot, a miner, a forester, a kid walking home from school. We know the next story could be the one that cracks the silence.

Or not. That's the gamble.

But that's what makes it real.

Every time we head back into the bush, we carry the stories of those who came before us. The ones who cast the first tracks. Who stayed up listening to knocks in the dark. Who documented what others laughed off. This search isn't a trend. It's a lineage.

And for all we don't know—for all we *still* haven't proven after seventy years—here's what we do have:

We have a continent of patterns.
We have thousands of overlapping reports.
We have tracks, and howls, and whispers in the treeline.

We have one another.
We have reason to keep going.

And most of all—we have a question that's never stopped asking back.

So what comes next?

Another ridge.
Another camera.
Another long night of waiting.

And maybe—just maybe—an answer.

But even if it never comes, we'll still be out there.

Because the moment we stop looking… that's when the mystery dies.

And some things are worth keeping alive.

Appendix I — Terminology of the Field: A Researcher's Glossary

Class A Sighting — A direct, unobstructed visual encounter with a Bigfoot-like creature under clear conditions. The gold standard for credible reports.

Class B Sighting — Indirect evidence such as movement through trees, vocalizations, or shadows—where a creature was sensed but not clearly seen.

Trackway — A consecutive line of footprints, usually in soft ground, snow, or sand. Often used to measure stride, gate, and pressure depth.

Vocalization — Any recorded or reported sound attributed to the creature, including howls, whoops, screams, chatter, or "samurai speech."

Wood Knock — A sharp knocking sound, often in pairs or threes, believed to be used for communication. Made by striking wood against wood.

Gifting — A behavior some researchers claim involves the creature leaving objects (e.g. stones, feathers) or accepting offered food. You do not subscribe to this belief.

Infrasound — Extremely low-frequency sound waves, below human hearing range, believed by some to cause nausea or confusion. A controversial but recurring topic.

Silence Zone — A phenomenon where forest noise abruptly stops—no birds, no bugs, no wind. Often associated with nearby animal presence or sightings.

Tree Structures — Unnaturally arranged limbs, arches, or stacked branches found deep in forests. Some believe they serve as markers or shelters.

Bluff Charge — Aggressive but non-contact behavior. Fast approach followed by retreat, reported in both bears and possible Sasquatch encounters.

Stick Sign Language — Hypothetical visual communication using stick patterns. Largely speculative and not widely accepted among data-focused researchers.

Habituation — The idea that a Bigfoot may become used to human presence and visit a location regularly. Often misused in anecdotal claims.

Sighting Corridor — A geographic region where multiple credible sightings and evidence clusters along natural travel routes like rivers or ridgelines.

Thermal Signature — Heat-based imagery from thermal cameras. Often used in night research to detect warm-bodied movement beyond visible light range.

Break Line — A sequence of fresh tree breaks at shoulder height, typically found in remote areas. Some believe it indicates movement or territory.

Field Cast — A plaster or silicone mold taken of a footprint, handprint, or impression believed to be left by a Sasquatch.

First Nations Knowledge — Oral histories from Indigenous communities, often including stories of wildmen or bush beings. Provides cultural context, not always literal sightings.

Bipedal Gait — A two-footed walking pattern. Sasquatch reports often note an unusual, fluid stride different from human mechanics.

Sighting Dead Zone — An area where credible reports suddenly stop, often after clearcutting, urban development, or increased traffic.

Researcher Bias — The influence of personal belief or desire for proof on interpretation of ambiguous evidence. A challenge for both skeptics and believers.

Appendix II — Misidentification Hazards: Fauna and Vocal Confusion

Black Bears (Ursus americanus) — The most commonly misidentified animal in North America. Capable of standing upright, especially when startled or reaching. Their silhouette can resemble a biped in low light. Cub behavior and adult aggression can mimic "creature" sightings, especially near food sources.

Moose (Alces alces) — Extremely large and often dark, a moose moving through dense brush can appear humanoid at a glance. In snow or dusk, antlers may be hidden, adding to confusion.

Humans in Camouflage — Hunters or loggers wearing ghillie suits, hoods, or full camo gear can appear creature-like, particularly at a distance. Nighttime movement and silence may contribute to a false sighting.

Barred Owls (Strix varia) — Their vocalizations often sound like someone screaming, yelling "who cooks for you?" or producing guttural howls. Frequently mistaken for primate-like sounds in the woods.

Foxes (Vulpes vulpes) — Fox screams are unearthly and blood-curdling. These cries are often reported as "woman screaming in woods" and misattributed to cryptids.

Ravens and Crows — These birds can mimic human voices and odd sounds. In fog or dense bush, their vocalizations can echo and appear directionless.

Porcupine Chatter — The high-pitched, rhythmic teeth chattering or squealing can sound unnatural and eerie when echoing through timber stands.

Wind Distortion — Natural wind passing through hollow logs, rock crevices, or canyons can create harmonics mistaken for moaning or howls.

Feral Dogs or Coyotes — Packs of coyotes yipping and howling may mimic primate calls when bouncing off valley walls or lake surfaces.

Bear Scat and Odor — Black bears can emit an intense musky smell, especially in mating season or during territorial displays. This odor is sometimes described identically to the "foul smell" linked to Sasquatch encounters.

Tree Shadows and Stump Figures — Pareidolia (the brain's tendency to see faces/forms) often causes people to misidentify tree stumps, shadow play, or root formations as upright figures.

Blowdowns and Rootballs — Fallen trees with exposed root systems can resemble massive creatures hunched or seated when partially obscured.

Boot Prints and Heel Overlap — In snow or soft mud, overlapping boot prints or partial human impressions can resemble large bare feet.

Logging Activity or Machinery Echoes — Distance and terrain can distort mechanical sounds, making them seem animal-like or vocal.

Bear Charging Behavior — Bears may "bluff charge" by running on all fours and rearing up. This rapid transition from horizontal to vertical can cause panic and misidentification.

Smoky or Low-Light Conditions — Low visibility often leads witnesses to overestimate height, shape, and features. Most misidentifications occur within 90 seconds of initial visual contact.

Appendix III — Field Equipment in 2025: What Works, What Fails

Trail Cameras (Infrared vs. Thermal) — Traditional motion-activated infrared cams are reliable for wildlife but often fail to capture Sasquatch due to avoidance behavior, low placement, or scent detection. Thermal cams provide heat-based imaging, useful at night and in winter, but are expensive and require careful angling.

Parabolic Microphones — Highly sensitive directional mics that can capture distant knocks, howls, or movement. Effective in still environments but vulnerable to wind distortion. Best when paired with real-time monitoring.

Audio Recorders (Long-Duration) — Compact digital units like the Zoom H2n or Tascam DR-series can run for hours or days, especially with battery packs. Used to collect overnight forest audio in suspected corridors.

Thermal Scopes and Monoculars — Handheld devices that detect heat signatures. Excellent for observing at a distance, but identifying features remains difficult without follow-up.

Drones (Thermal-Equipped) — Increasingly common in 2025, these units offer short-range thermal scanning of hard-to-reach terrain. Limited by battery life, weather, and flight noise.

GPS Units and Trackers — Essential for remote navigation and for tagging sighting locations. Some models allow field notes or image linking. Preferred over phone-based GPS in deep wilderness zones.

Headlamps with Red Light Mode — Crucial for nighttime travel without alerting wildlife. Red light helps preserve night vision and is less likely to spook animals—or anything else.

Field Notebooks (Weatherproof) — Waterproof paper and pens allow for immediate documentation even in rain or snow. Notes taken at the moment are vastly more reliable than later recollection.

Casting Kits (Plaster or Silicone) — Used for preserving prints. Plaster is affordable and field-friendly but can crack; silicone offers more precision and durability but is pricier.

Cell Phones (Limited Usefulness) — While useful for mapping and quick video, phone-based tools are hindered by poor reception, weak microphones, and battery drain in cold weather.

Personal Locator Beacons (PLBs) — Emergency tools that can transmit a distress signal via satellite. Essential for solo or deep-wilderness research.

Camouflage and Scent Control Gear — Neutral clothing, scent-masking sprays, and scent-free soaps are sometimes used to avoid detection. Effectiveness is debated in Bigfoot-specific research.

Firearms and Bear Spray — Carried for protection in bear country, not for engagement. Use of weapons is controversial and must follow all safety and legal regulations.

Immediate Note-Taking — Write down everything as soon as possible, while memory is fresh. Use field notebooks or voice recorders. Record time, date, location, weather, lighting, and anything unusual (odors, animal behavior, environmental silence).

Photographic Evidence — Take multiple photos from various angles. Include objects for scale: a boot, glove, or a measuring tape next to tracks or markings. Avoid zooming—move closer when possible.

Footprint Casting — Use plaster or silicone. Clear debris carefully, then pour slowly into the print. Let cure fully before removal. Mark the cast with date and location. Include GPS coordinates if available.

Audio Recording — Keep a long-running audio device active in suspected areas. Mark exact time of unusual sounds. If vocalizations are heard, stay silent and let the recorder do the work.

GPS Pinpointing — Use handheld GPS or apps to log coordinates of evidence. Mark trail access points, campsite locations, and direction of creature movement.

Sketches and Maps — Even a rough sketch of the scene can help later analysis. Indicate witness location, movement direction, physical features, and distances.

Environmental Context — Record temperature, wind, moon phase, and animal activity before and after the encounter. Note any silences or shifts in behavior among birds and mammals.

Witness Statements — Record interviews as soon as possible. Ask open-ended questions. Avoid leading phrasing. Let the witness tell the story in their own words.

Avoiding Contamination — Do not step in, around, or over prints. Do not handle biological samples without gloves or sterile tools. Mark the area with flags or tape if available.

Follow-Up Timeline — Return to the site as soon as safely possible for additional inspection, casting, scent dogs, or second opinions. Track degradation of evidence over time.

Reporting — Submit evidence and reports to a credible database or researcher group. Include your contact information, method of discovery, and any supporting files.

Appendix V — Credible Voices: Researchers, Institutions, and Resources

Dr. Jeffrey Meldrum (RIP)— Professor of anatomy and anthropology at Idaho State University. Known for his extensive cast collection (over 300) and scientific approach to footprint morphology. One of the few academics openly investigating the phenomenon.

Grover Krantz (1931–2002) — Physical anthropologist from Washington State University. Early academic advocate for Bigfoot research. His work laid foundational standards for evaluating footprint casts and gait analysis.

John Green (1927–2016) — Canadian journalist and author of several foundational Bigfoot books. His collected reports helped establish patterns in North American sightings.

Dr. John Bindernagel (1941–2018) — Canadian wildlife biologist and author of *The Discovery of the Sasquatch*. Advocated for the animal's acceptance as a North American species, emphasizing field evidence and ecological logic. One of the few researchers with professional credentials in wildlife biology who actively pursued the topic.

Ontario-Bigfoot.com — One of the most regionally relevant and active platforms for Bigfoot reports in Ontario. Hosts dozens of firsthand accounts, blog, video's, photos, and regional encounter data.

The North American Wood Ape Conservancy (NAWAC) — A group focused on evidence-based field research in the southern U.S. Operates long-term observation areas and uses military-style field methods.

The Bigfoot Field Researchers Organization (BFRO) — One of the largest databases of North American sighting reports, searchable by region and year. Reports are vetted and categorized by credibility.

Books: "Sasquatch: Legend Meets Science" by Jeff Meldrum — A deep dive into biomechanics, dermal ridges, and physical evidence. Balances scientific inquiry with open-minded exploration.

Books: "The Discovery of the Sasquatch" by John Bindernagel — Written by a wildlife biologist, this book argues for acceptance of Sasquatch as a North American species based on ecological reasoning.

Podcasts: "Sasquatch Chronicles," "Bigfoot Eyewitness Radio," "From the Shadows" — Popular audio formats where witnesses share personal encounters in detail. Some include commentary from researchers.

Documentaries: "On the Trail of Bigfoot" (Small Town Monsters), "Sasquatch: Legend Meets Science," "Bigfoot: The Definitive Guide" — Credible video productions featuring physical evidence, historical context, and firsthand investigation footage.

Appendix VI — Safety and Ethics in Wilderness Field Work

Respect Private and Indigenous Lands — Always know where you are. Use accurate GPS maps and land use data to avoid trespassing. Gain permission before entering Indigenous or privately held territory. Understand that these areas may contain sacred spaces or protected zones. Respecting these boundaries builds trust and protects access.

Travel With a Partner or Notify Someone — The Canadian wilderness is unforgiving. Always inform someone of your route, expected return time, and access points. Lone travel in remote zones should be avoided unless you're experienced, well-equipped, and in radio or satellite contact.

Carry Appropriate Protection — This may include bear spray, signal flares, or—where legal and trained—firearms. Know the laws in your province or state. Use deterrents, not threats. Never enter the woods assuming you'll need to defend yourself from a Sasquatch. Your biggest danger is environmental: terrain, weather, and large predators.

Pack Redundant Gear — Bring backup power sources, thermal blankets, extra batteries, waterproof fire starters, and backup nav tools. Cold, wet, and darkness have ended more expeditions than any creature encounter.

Don't Chase or Harass Wildlife — Whether bear, moose, or something unknown, never pursue. Record from a distance. Let the evidence come to you. Too many amateur researchers ruin potential evidence—or get injured—by trying to "get closer."

Practice Quiet Observation — Sit, watch, and wait. Movement and talking reduce your odds of seeing wildlife. Let the forest settle before expecting activity. Move slowly and intentionally. Use scent-free soaps, avoid food odors, and mute gear where possible.

Avoid Speculative Behavior — Do not scream, bang trees, or throw rocks. These behaviors have never produced evidence under scientific conditions and often scare away natural wildlife or mimic aggression.

Leave No Trace — Every researcher has a duty to preserve the bush. Pack out everything. Do not cut live trees, alter environments, or leave bait. Baiting is both ethically questionable and ineffective based on current evidence.

Handle Evidence with Integrity — Never stage, exaggerate, or "enhance" a find. The credibility of this field suffers every time someone manipulates data for attention. Let the truth be strange enough.

Keep an Open but Critical Mindset — Curiosity is essential, but so is skepticism. Be your own harshest critic. Double-check every sound, every print, every story. Treat your research like it will be peer-reviewed—because someday it might be.

Know When to Turn Back — Weather shifts, flooding, fatigue, or gear failure are signs to call it. Pride should never outweigh safety. A real field researcher knows survival comes first. Always.

www.ingramcontent.com/pod-product-compliance
Lightning Source LLC
Chambersburg PA
CBHW080331270326
41927CB00014B/3174